Cosmetics Ingredients Research

Cosmetics Ingredients Research

Guest Editor

Anna Waszkielewicz

MDPI

Basel • Beijing • Wuhan • Barcelona • Belgrade • Novi Sad • Cluj • Manchester

Guest Editor
Anna Waszkielewicz
Department of Medicinal
Chemistry and Biochemistry
Andrzej Frycz Modrzewski
Krakow University
Krakow
Poland

Editorial Office
MDPI AG
Grosspeteranlage 5
4052 Basel, Switzerland

This is a reprint of the Special Issue, published open access by the journal *Applied Sciences* (ISSN 2076-3417), freely accessible at: https://www.mdpi.com/journal/applsci/special_issues/6GCEZ3H494.

For citation purposes, cite each article independently as indicated on the article page online and as indicated below:

Lastname, A.A.; Lastname, B.B. Article Title. *Journal Name* **Year**, *Volume Number*, Page Range.

ISBN 978-3-7258-5929-0 (Hbk)
ISBN 978-3-7258-5930-6 (PDF)
https://doi.org/10.3390/books978-3-7258-5930-6

Contents

About the Editor

Anna Waszkielewicz

Anna Waszkielewicz is a doctor of pharmaceutical science at the Department of Medicinal Chemistry and Biochemistry, Faculty of Medicine, Medical College, Andrzej Frycz Modrzewski Krakow University in Krakow. She is the co-author of 90 publications, with a total impact factor of 187,263, in the field of medicinal chemistry and cosmetology. She is the editor of the Special Issues "Development of Innovative Cosmetics" and "Cosmetics Ingredients Research" in the journal *Applied Sciences*. She is a lecturer and coordinator of cosmetics chemistry and organic chemistry at the Faculty of Pharmacy at the Jagiellonian University, and was formerly a lecturer in Perugia, Freiburg and Würzburg in the field of cosmetics and organic chemistry. She has supervised doctoral theses in the field of medicinal chemistry, including the chemistry of substances used in cosmetics.

applied sciences

MDPI

Editorial

Peptides and Their Mechanisms of Action in the Skin †

Anna Maria Waszkielewicz [1,*] and Kaja Mirosław [2]

1 Department of Bioorganic Chemistry, Chair of Organic Chemistry, Faculty of Pharmacy, Jagiellonian University Medical College, Medyczna 9, 30-688 Krakow, Poland
2 Institute of Quality Sciences and Product Management, Cracow University of Economics, Rakowicka 27, 31-510 Krakow, Poland; kajamiroslaw@gmail.com
* Correspondence: anna.waszkielewicz@uj.edu.pl; Tel.:+48-12-6205594

† I dedicate this editorial to my cosmetology students. Anna Waszkielewicz

1. Introduction

The skin, the largest organ of the human body, not only has a protective function, but also plays a key role in aesthetics and health. It is a physical barrier separating us from the outside world, but at the same time a living tissue in which complex biological processes constantly take place [1]. Over the years, under the influence of internal and external factors, such as UV radiation, from which it protects us thanks to pigment cells, oxidative stress or hormonal changes as well as gravity, the skin loses its elasticity, firmness and ability to regenerate [2]. Natural skin aging contributes to reduced production and increased degradation of extracellular matrix proteins, such as collagen, fibronectin, elastin and laminin, which are the frame for skin cells [3]. The aging process, inevitable and visible, is becoming a challenge that both scientists and cosmetology specialists struggle with. It is, however, a motivation in the search for solutions for maintaining the youthful appearance of the skin, where more and more attention is being paid to modern active ingredients such as bioactive and biomimetic peptides, which offer unique possibilities of supporting natural repair processes and counteracting the signs of aging [4].

2. An Overview of Publications and Intellectual Property Literature

Unique properties of peptides are valued due to their participation in the regulation of homeostasis, stress, immunity, defense against pathogens and regeneration [5]. In terms of cosmetological use, the advantages of peptides include non-invasiveness, possibility of self-application (no injection) and higher safety comparing to botox. On the other hand, their disadvantages are the time needed for visible effect measured in weeks (except for Syn-Ake), lability of structure, and of course, price resulting from manufacturing and storage conditions.

Bioactive peptides, usually consisting of 3–30 amino acids, come from natural proteins and act as signaling molecules or substrates involved in various biological processes. They are short chains of amino acids (up to 50), connected by peptide bonds, which distinguishes them from proteins (100 or more amino acids) [4,6]. Peptides occur in our body and are produced by various tissues, fulfilling regulatory, transporting, signaling, immune response, hormone and many other roles, but some of the peptides called biomimetic are produced synthetically [7].

Peptides can also be classified according to the division into signal peptides, neurotransmitter peptides, enzyme inhibitor peptides and carrier peptides [6].

2.1. Signal Peptides

Signal peptides are active substances that can counteract the skin aging process by stimulating fibroblasts to act. As a result, there is an increased biological response, such as increased production of collagen, elastin, fibronectin, glycosaminoglycans, and proteoglycans. These peptides can also act as growth factors by activating protein kinase C, which plays a key role in cell growth and migration [3]. Signal peptides are used either to enhance metabolism of fibroblasts or fasten hair growth.

Carnosine (*β-(Ala–His)*) is a dipeptide which is also used as a food supplement. Its wound healing activity is a feature making it a cosmetic, especially for eye cream, because *stratum corneum* around the eyes is the thinnest [8]. It has been bound with palmitic acid into palmitoyl-carnosine in order to enhance permeation through *stratum corneum* so that the compound can be effectively used on the whole face (Figure 1).

Figure 1. Chemical structure of palmitoyl-carnosine. Palmitoyl group is marked red.

Palmitoyl tripeptide-3 (Syn-Coll, (*Pal–Lys–Val–Lys*) bistrifluoracetate) stimulates synthesis of collagen, also through TGF-β [9].

A palmitoylated peptide is Pal-KTTKS (Matrixyl, palmitoyl pentapeptide)—Lys-Thr-Thr-Lys-Ser. KTTKS is a procollagen 1 fragment and it stimulates synthesis of collagen, elastin and hyaluronic acid [10]. Another peptide, also patented by Sederma, Pal-GQPR (palmitoyl 7-tetrapeptide, Matrixyl 3000)—Gly-Glu-Pro-Arg, stimulates collagen synthesis by fibroblasts. Moreover, it inhibits production of interleukin IL-6 resulting in anti-inflammatory effect. $Pal - KMO_2K$ under the tradename MATRIXYL Synthe-6 (MO_2 corresponding to a dioxygenated methionine) [9].

VGVAPG constitutes a fragment of elastin (elastin-derived peptides EDPs) and is a chemotactic signal peptide for fibroblasts. EDPs initiate human mesenchymal stem cells (hMSCs) differentiation into more tissue-specific cells [11]. It also decreases in the level of reactive oxygen species (ROS) [12].

Certain peptides have been developed for stimulation of hair growth. For example, tripeptide-1 combined with biotin (vitamin H)—biotinyl GHK in Lash Accelerator—or a preparation preventing hair loss Procapil. The longest of this group is KGF (Keratin Growth Factor) used in EyeLash Enhancer, containing 164 amino acids.

Leuphasyl—pentapeptide-18 (*Tyr–D–Ala–Gly–Phe–Leu*) inhibits secretion of sebum. A response for androgenic hair loss may be acetyl-tetrapeptide-3 [13]. It also decreases neuronal excitation by imitating activity of enkefalins [14].

The tradename Renokin, includes decapeptide-10, oligopeptide-54 (CG-Nokkin), decapeptide-18, acetyl decapeptide-3, and oligopeptide-42 which promote hair growth [15].

There are also peptides preventing hair greying: α-melanocyte stimulating hormone (acetylhexapeptide-1, *Ac–Nle–Ala–His–D-Phe–Arg–Trp–*NH_2) [16,17] Melitane)-α-MSH

biomimetic peptide. On the other hand, PTP-20 (Palmitoyl tetrapeptide-20, Greyverse) was developed for the use of MC1 receptor agonist and it increases quantity of melanin in human melanocyte culture [18].

Aquashine PTX 4 contains decapeptide-29, oligopeptide-62, acetyl decapeptide-3, oligopeptide-24, oligopeptide-72, oligopeptide-34, oligopeptide-51 with sodium hyaluronate [19]—the peptide ingredients of the preparation have been patented for all possible uses in the skin. *Ac–Asp–Val–Lys–Tyr–OH* increases the expression of fibulin-5, increases synthesis of type I collagen as well as elastin through overexpression of various genes [20].

2.2. Enzyme Inhibitors

An example of peptides that are direct and indirect enzyme inhibitors are rice and soy peptides. Soy peptides obtained from soy are proteinase inhibitors [3]. Rice peptides, on the other hand, are natural proteins that inhibit the action of matrix metalloproteinases and stimulate the expression of the gene responsible for the synthesis of hyaluronan 2 in keratinocytes [21].

2.3. Neurotransmitter Peptides

An example of a peptide that inhibits neurotransmitters is acetyl hexapeptide-8 (Argireline, $Ac–Glu–Glu–Met–Gln–Arg–Arg–NH_2$) (Figure 2), which has an effect similar to botulinum toxin. It inhibits docking of vesicles in neuromuscular synapses, which reduces secretion of acetylcholine to the synapse and as a consequence, the formation of wrinkles [19]. The structure was based on SNAP-25 protein. It reduces wrinkles, however, the effect is visible after nearly a month [22].

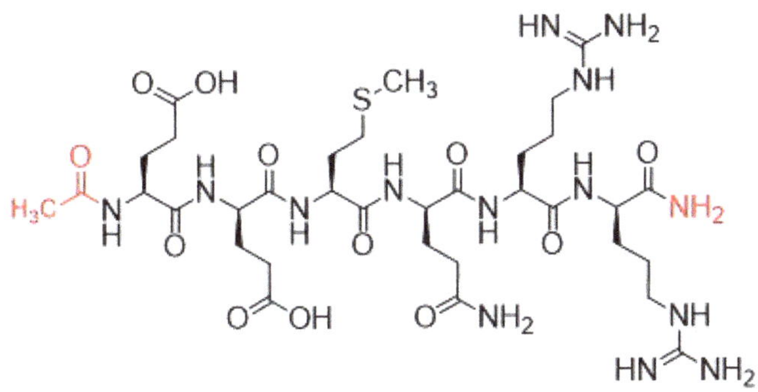

Figure 2. Chemical structure of argireline. Acetyl and amide groups are marked red.

SNAP-8—acetyl-octapeptide-3 $Ac - Glu - Glu - Met - Gln - Arg - Ala - Asp - NH_2$ (acetyl-glutamyl-heptapeptide-3) has the same mechanism of action as argireline [23].

Syn-Ake (β-Ala-Pro-GABA-benzylamide) represents reversible antagonism of N_m receptor and was designed on the basis of waglerin-1, temple pit viper (Tropidolaemus wagleri) venom. Waglerin-1 contains 22 amino acids and obviously is too large to get easily absorbed through stratum corneum. Syn-Ake has an immediate antiwrinkle effect and can be used directly before a party. On the other hand, this peptide is also added to cosmetics containing Argireline in order to fasten visibility of effects. Pentapeptide-3 (Vialox, Gly-Pro-Arg-Pro-Ala-NH2) is a competitive Nm antagonist. Moreover, is has sebostatic activity [14].

2.4. Carrier Peptides

The last type of peptide is carrier peptides, and in this case we will look at the copper peptide hidden in cosmetology under the name Copper Tripeptide-1. Gly-His-Lys (GHK, matricryptin) is the biological name of the tripeptide isolated from human plasma, showing a strong affinity for copper (II) ions, with which it forms a complex (GHK-Cu) spontaneously. Initially, it was identified as a growth factor for various mature cells, but more recent studies indicate its physiological role in wound healing and tissue regeneration. GHK-Cu directly supports wound healing and stimulates biological processes essential for tissue repair, such as angiogenesis, nerve development, and chemoattraction of key cells for the healing process (e.g. macrophages, monocytes, mast cells, capillary endothelial cells) [24]. Maquart et al. reported that GHK-Cu stimulates collagen synthesis in cultured fibroblasts, independent of the action of growth factors [25]. The presence of the GHK sequence in the $\alpha 1$(I) chain of human collagen suggests that this tripeptide may be released during the collagen production process. It is also very helpful in the wound healing process, where it supplies copper for matalloproteinase, hence its classification as a carrier peptide [6]. The copper peptide was initially confused with a growth factor, but Pickart et al. demonstrated that it has a stimulating effect on fibroblasts [26,27]. It reduces reactive oxygen species (ROS) production, increases superoxide dismutase (SOD) activity while decreases TNF-α and IL-6 production through the suppression of NF-κ B, p65 and p38 MAPK signaling [28]. The tripeptide has been also subject to modification. Pal-GHK stimulates fibrillogenesis probably by activity on TGF-β [8]. Biotinyl-GHK is also capable of binding copper [29].

The proposed mechanisms of action refer to many compounds, and there are many more peptides, contained in preparations under registered names, as listed below, in alphabetic order [9,19]:

1. ActiMatrix: Peptide based mushroom Extract;
2. Actimp 1.9.3: Hydrolyzed Lupine Protein
3. Adifyline: Acetyl Hexapeptide-38
4. Aldenine: Hydrolized Wheat Protein, Hydrolized Soy Protein, Tripeptide- 1
5. Ascotide: Ascorbyl Phosphate Succinoyl Pentapeptide-12
6. BeauActive MTP: Hydrolyzed milk protein
7. Bio-Bustyl: Rahnella Soy Protein Ferment, Palmitoyl Oligopeptide
8. Biopeptide CL: Palmitoyl Oligopeptide
9. Biopeptide EL: Palmitoyl Oligopeptide
10. BONT-L-Peptide: Palmitoyl Hexapeptide-19
11. Caspaline 14: Hexapeptide-42
12. ChroNOgen: Tetrapeptide-26
13. ChroNOline: Caproyl Tetrapeptide-3
14. Colhibin: Hydrolyzed Rice Protein
15. Collaxyl: Hexapeptide-9
16. Cytokino: Hydrolyzed Casein, Hydrolyzed Yeast Protein
17. Decorinol: Tripeptide-9 Citrulline
18. Decorinyl: Tripeptide-10 Citrulline
19. Deepaline: Palmitoyl hydrolyzed Wheat Protein
20. Delisens: Acetyl Hexapeptide-46
21. DermaPep A350: Myristol Tripeptide-31
22. DermaPep A420: Myristoyl Tetrapeptide-6
23. Dermican: Acetyl Tetrapeptide-9
24. dGlyage: Lysine HCl, Lecithin, Tripeptide-9 Citrulline
25. Diffuporine: Acetyl Hexapeptide-37
26. Drieline: Yeast Betaglucan
27. Dynachondrine ISR: Hydrolized Soy Protein
28. ECM Moduline: Palmitoyl Tripeptide-28
29. ECM Protect: Tripeptide-2
30. Effipulp: Hydrolyzed Avocado Protein
31. Elhibin: Glycine Soja Protein
32. Extracellium: Hydrolyzed Potato Protein
33. Eyeseryl: Acetyl Tetrapeptide-5
34. Granactive AGE: Palmitoyl Hexapeptide-14
35. Inyline: Acetyl Hexapeptide-30
36. IP 2000: Trifluoroacetyl Tripeptide-2
37. Juvefoxo: Acetyl Hexapeptide-50
38. Kollaren: Tripeptide-1
39. Laminixyl IS: Heptapeptide
40. Leuphasyl: Pentapeptide-18
41. Liftline: Hydrolyzed Wheat Protein
42. Lipacide PVB: Palmitoyl hydrolyzed Wheat Protein
43. Lipeptide: Hydrolized Vegetable Protein
44. Marine Filling Spheres: Atelocollagen
45. Pentacare-NA: Hydrolyzed Wheat Gluten
46. Pep 4-17: Tetrapeptide-17

47. Pepha-Timp: Human oligopeptide-20
48. Peptamide 6: Hexapeptide-11
49. Peptide AC29: Acetyl Tripeptide-30 Citrulline
50. Peptide Q10: Pentapeptide-34 Trifluoroacetate
51. Peptide Vinci 01: Penta-decapeptide-1
52. Peptide Vinci 02: Hexapeptide-3
53. Peptiskin: Arginine/Lysine polypeptide
54. Preregen: Glycine soja (Soybean) Protein, Oxido Reductases
55. Preventhelia: Diaminopropionoyl Tripeptide-33
56. Prodejine: Yeast Extract
57. Prolixir S20: Dimer Tripeptide-43
58. Quintescine: Dipeptide-4
59. Raffermine: Hydrolyzed Soy Flour
60. Relistase: Acetylarginyltriptophyl Diphenylglycine
61. Renaissance: Hydrolyzed Wheat Protein, Palmitoyl Decapeptide-21, Decapeptide-22, Oligopeptide-78, Zinc Palmitoyl Nonapeptide-14
62. Regu-Age: Hydrolyzed Rice Bran Protein, Glycine Soja Protein
63. Ridulisse C: Hydrolyzed Soy Protein
64. Serilesine: Hexapeptide-10
65. Silusyne: Soybean (Glycine Soja) Oil, Lauryldimonium Hydroxypropyl Hydrolized Soy Protein, Acetyl Hexapeptide-39
66. SNAP-7: Acetyl Heptapeptide-4
67. Survixyl IS: Pentapeptide- 31
68. Syn-Glycan: Tetradecyl Aminobutyroylvalyl-aminobutyric Urea Trifluoroacetate
69. Syn-Tacks: Palmitoyl Dipeptide-5 Diaminobutyloyl Hydroxythreonine, Palmitoyl Dipeptide-6 Diaminohydroxybutyrate
70. Syn-TC: Tetradecyl Aminobutyroylvalylaminobutyric Urea Trifluoroacetate, Palmitoyl Tripeptide-5, Palmitoyl Dipeptide-5 Diaminobutyroyl Hydroxythreonine
71. Syniorage: Acetyl Tetrapeptide-11
72. TEGO Pep 4-Even: Tetrapeptide-30
73. Telangyn: Acetyl Tetrapeptide-33
74. Telosense: Hydrolized Soy Protein, Hydrolized Yeast Protein
75. Thermostressine: Acetyl Tetrapeptide-22
76. Thymulen 4: Acetyl Tetrapeptide-2
77. TIMP Peptide: Acetylhexapeptide-20
78. Triactigen: Yeast Extract
79. Trylagen: Hydrolyzed Wheat Protein, Hydrolyzed Soy Protein, Tripeptide-10 Citrulline, Tripeptide-1

3. Conclusions

This review presents variability of mechanisms of action of various peptides used in cosmetology. It is worth mentioning that law referring to cosmetics ingredients is not as much restrictive as pharmaceutical law, therefore, we do not really know all the mechanisms, and disadvantages, as well as long term toxicity of these compounds. A few years ago it looked like these few compounds that have been described will be enough on the market, but patent laws encouraged cosmetology producers to search for new active peptides for which they may have their own patent and resulting from it monopoly. It is worth observing how this domain of cosmetological peptides evolves.

Author Contributions: Conceptualization, A.M.W.; investigation, A.M.W. and K.M.; resources, A.M.W. and K.M.; writing—original draft preparation, K.M. and A.M.W.; writing—review and editing, A.M.W.; visualization, A.M.W.; supervision, A.M.W. All authors have read and agreed to the published version of the manuscript.

Funding: This research received no external funding.

Conflicts of Interest: The authors declare no conflicts of interest.

References

1. Rawlings, A.V.; Harding, C.R. Moisturization and Skin Barrier Function. *Dermatol. Ther.* **2004**, *17*, 43–48. [CrossRef] [PubMed]
2. Tobin, D.J. Biochemistry of Human Skin—Our Brain on the Outside. *Chem. Soc. Rev.* **2006**, *35*, 52–67. [CrossRef] [PubMed]
3. Pai, V.; Bhandari, P.; Shukla, P. Topical Peptides as Cosmeceuticals. *Indian J. Dermatol. Venereol. Leprol.* **2017**, *83*, 9–18. [CrossRef] [PubMed]
4. He, B.; Wang, F.; Qu, L. Role of Peptide—Cell Surface Interactions in Cosmetic Peptide Application. *Front. Pharmacol.* **2023**, *14*, 1267765. [CrossRef]
5. Usman, R.; Bharadvaja, N. Nutricosmetics: Role in Health, Nutrition, and Cosmetics. *Proc. Indian Natl. Sci. Acad.* **2023**, *89*, 584–599. [CrossRef]
6. Ngoc, L.T.N.; Moon, J.Y.; Lee, Y.C. Insights into Bioactive Peptides in Cosmetics. *Cosmetics* **2023**, *10*, 111. [CrossRef]
7. Lima, T.N.; Moraes, C.A.P. Bioactive Peptides: Applications and Relevance for Cosmeceuticals. *Cosmetics* **2018**, *5*, 21. [CrossRef]
8. Schagen, S.K. Topical Peptide Treatments with Effective Anti-Aging Results. *Cosmetics* **2017**, *4*, 16. [CrossRef]
9. Peschard, O.; Doucet, A.; Leroux, R.; Mondon, P. Peptides, Compositions Comprising Them and Uses in Particular Cosmetic Uses 2014. EP 3 145 943 B1, 21 May 2015.
10. Tałałaj, U.; Uścinowicz, P.; Bruzgo, I.; Surażyński, A.; Zaręba, I.; Markowska, A. The Effects of a Novel Series of KTTKS Analogues on Cytotoxicity and Proteolytic Activity. *Molecules* **2019**, *24*, 3698. [CrossRef]
11. Szychowski, K.A.; Skóra, B. Elastin-Derived Peptides (EDPs) Affect Gene and Protein Expression in Human Mesenchymal Stem Cells (HMSCs)—Preliminary Study. *Cytokine* **2024**, *182*, 156725. [CrossRef]
12. Skóra, B.; Piechowiak, T.; Szychowski, K.A. Interaction Between Aging-Related Elastin-Derived Peptide (VGVAPG) and Sirtuin 2 and Its Impact on Functions of Human Neuron Cells in an In Vitro Model. *Mol. Neurobiol.* **2024**, *Online ahead of print*. [CrossRef]
13. Carreno Serraima, C.; Garcia Sanz, N.; Cebrian Puche, J.; Alminana Domenech, N.; Ferrer Montiel, A.; Van Den Nest, W. Useful Peptides in the Treatmnet and/or Care of Skin, Mucouses and/or Leather and Its Use in Cosmetic or Pharmaceutical Compositions. EP 2 349 972, 23 October 2009.
14. Zhmak, M.N.; Utkin, Y.N.; Andreeva, T.V.; Kudryavtsev, D.S.; Kryudova, E.V.; Tsetlin, V.I.; Shelukhina, I.V.E. Peptide Inhibitors of Nicotinic Acetylcholine Receptor. US 9550,808 B2, 20 January 2014.
15. Waugh, J. Compositions for Topical Application of Compounds. US 2017/0290778 A1, 14 September 2018.
16. Ye, Y.; Li, Y.; Xu, C.; Wei, X. Improvement of Mild Photoaged Facial Skin in Middle-Aged Chinese Females by a Supramolecular Retinol plus Acetyl Hexapeptide-1 Containing Essence. *Ski. Health Dis.* **2023**, *3*, 1–9. [CrossRef] [PubMed]
17. Boo, Y.C. Intrinsic and Extrinsic Factors Associated with Hair Graying (Canities) and Therapeutic Potential of Plant Extracts and Phytochemicals. *Appl. Sci.* **2024**, *14*, 7450. [CrossRef]
18. Almeida Scalvino, S.; Chapelle, A.; Hajem, N.; Lati, E.; Gasser, P.; Choulot, J.C.; Michel, L.; Hocquaux, M.; Loing, E.; Attia, J.; et al. Efficacy of an Agonist of α-MSH, the Palmitoyl Tetrapeptide-20, in Hair Pigmentation. *Int. J. Cosmet. Sci.* **2018**, *40*, 516–524. [CrossRef] [PubMed]
19. Ferrer Montiel, A.V.; Alminana Domenech, N.; Cebrian Puche, J.; Van Den Nest, W.; Carreno Serraima, C.; Delgado Gonzalez, R. Compounds Useful in the Treatment and/or Care of the Skin and Their Cosmetic or Pharmaceutical Compositions. EP 2 986 626 B1, 15 April 2014.
20. Wang, Y.; Wang, M.; Xiao, S.; Pan, P.; Li, P.; Huo, J. The Anti-Wrinkle Efficacy of Argireline, a Synthetic Hexapeptide, in Chinese Subjects: A Randomized, Placebo-Controlled Study. *Am. J. Clin. Dermatol.* **2013**, *14*, 147–153. [CrossRef]
21. Sim, G.S.; Lee, D.H.; Kim, J.H.; An, S.K.; Choe, T.B.; Kwon, T.J.; Pyo, H.B.; Lee, B.C. Black Rice (Oryza Sativa L. Var. Japonica) Hydrolyzed Peptides Induce Expression of Hyaluronan Synthase 2 Gene in HaCaT Keratinocytes. *J. Microbiol. Biotechnol.* **2007**, *17*, 271–279.
22. Blanes-Mira, C.; Clemente, J.; Jodas, G.; Gil, a; Fernández-Ballester, G.; Ponsati, B.; Gutierrez, L.; Pérez-Payá, E.; Ferrer-Montiel, a A Synthetic Hexapeptide (Argireline) with Antiwrinkle Activity. *Int. J. Cosmet. Sci.* **2002**, *24*, 303–310. [CrossRef]
23. Truniger, S; Wolfinger, T. Personal Care Composition. Manuf. Chem. US 11,278,475 B2, 5 April 2018.
24. Siméon, A.; Emonard, H.; Hornebeck, W.; Maquart, F. The Tripeptide-Copper Complex Glycyl-L-Histidyl-L-Lysine-Cu2+ Stimulates Matrix Metalloproteinase-2 Expression by Fibroblast Cultures. *Life Sci.* **2000**, *67*, 2257–2265. [CrossRef]
25. Maquart, F.X.; Pickart, L.; Laurent, M.; Gillery, P.; Monboisse, J.C.; Borel, J.P. Stimulation of Collagen Synthesis in Fibroblast Cultures by the Tripeptide-Copper Complex Glycyl-L-Histidyl-L-Lysine-Cu2+. *FEBS Lett.* **1988**, *238*, 343–346. [CrossRef]
26. Pickart, L.; Lovejoy, S. Biological Activity of Human Plasma Copper-Binding Growth Factor Glycyl-L-Histidyl-L-Lysine. *Methods Enzymol.* **1987**, *147*, 314–328. [CrossRef]
27. Pickart, L.; Vasquez-Soltero, J.M.; Margolina, A. GHK Peptide as a Natural Modulator of Multiple Cellular Pathways in Skin Regeneration. *Biomed Res. Int.* **2015**, *2015*, 648108. [CrossRef] [PubMed]
28. Park, J.R.; Lee, H.; Kim, S.I.; Yang, S.R. The Tri-Peptide GHK-Cu Complex Ameliorates Lipopolysaccharideinduced Acute Lung Injury in Mice. *Oncotarget* **2016**, *7*, 58405–58417. [CrossRef] [PubMed]
29. Tosto, R.; Vecchio, G.; Bellia, F. New Biotinylated GHK and Related Copper(II) Complex: Antioxidant and Antiglycant Properties In Vitro against Neurodegenerative Disorders. *Molecules* **2023**, *28*, 6724. [CrossRef] [PubMed]

applied sciences

MDPI

Article

A Study of the Potential of *Solidago virgaurea* Extract as a Raw Material for Cosmetic Macroemulsions

Marta Marzec [1,*], Grażyna Kaszczyk [2], Witold Sujka [2] and Izabela Nowak [1,*]

[1] Department of Applied Chemistry, Faculty of Chemistry, Adam Mickiewicz University, Poznań, Uniwersytetu Poznańskiego 8, 61-614 Poznań, Poland
[2] TRICOMED SA, Świętojańska 5/9, 93-493 Łódź, Poland
* Correspondence: marta.marzec@amu.edu.pl (M.M.); nowakiza@amu.edu.pl (I.N.); Tel.: +48-61-8291580 (I.N.)

Featured Application: The results proved that *Solidago virgaurea* extract can be concluded to be an active ingredient of key importance in inducing the anti-aging effect of the skin. This creates the possibility for cosmetic product manufacturers to use the studied plant extract as an ingredient in semi-solid cosmetic formulations with regenerative/anti-aging properties.

Abstract: The use of *Solidago virgaurea* extract as a raw material for cosmetics production fits the eco-tendency prevalent in the cosmetic industry for many years now perfectly. The study reported in this paper included an evaluation of the potential of the above-mentioned plant extract as a natural cosmetic ingredient applied in the form of cosmetic macroemulsions. The physicochemical parameters (pH, viscosity, particle size distribution) and physical stability (multiple light scattering) of the cosmetic formulations containing *Solidago virgaurea* (1.0 wt.%) were studied. An in vivo study was carried out on a group of 20 female volunteers. The results showed a high compatibility of the tested extract with the other components in the macroemulsion in the form of a serum for the body, which was the formulation containing Sorbitan Stearate as an emulsifier. Analysis of the results revealed a relation between the compatibility of the investigated herbal extract with the components of the cosmetic base and the effectiveness of this extract as an active substance. As a result of an improvement in the application parameters, an over 70% increase in the level of epidermis moisturization was observed along with other beneficial changes in the parameters of skin macrorelief and topography.

Keywords: emulsion; *Solidago virgaurea* extract; stability study; in vivo study

1. Introduction

The right choice of active ingredient is known to determine the character and effectiveness of a cosmetic product. For a few decades, much attention has been paid to replacing synthetic ingredients with natural ones. This interest in natural resources has brought about a dynamic development of the herbal processing industry that provides the producers of cosmetics with pure raw products and isolated plant extracts. The introduction of natural-origin raw products to cosmetic formulations has advantages as well as disadvantages. A great advantage is the possibility of using all parts of a given plant, the roots, stem, leaves and flowers, each of which contains a number of specific active components with beneficial effects on general health and the skin. The active substances contain a wide range of chemical compounds known to show a large variety of beneficial effects like antioxidative, anti-inflammatory, fragrance, dyeing, brightening, protecting against irradiation or moisturizing effects. From a technological viewpoint, the disadvantage of herbal raw products is the variation in the composition of the extracts from the same species of plant, which depends on the geographic site of origin and conditions of a given plant's growth and even on the time at which the plants were harvested [1]. One of the genera of herbal plants known to show therapeutic properties is *Solidago L.* The genus of *Solidago* originates from Asteraceae that are endemic to North America, Europe and Asia. In Poland, the most common species is *Solidago virgaurea* [2]. The plant contains many biologically active components [3], in particular, flavonoids (quercetin, rutoside, avicularin, kaempferol) [4], triterpene saponins, catechin tannins, phenol acids and glycosides [5–7]. An interesting component is the essential oil containing cyclic monoterpenes (α and β-pinene) [8]. The diversity of the active compounds is responsible for the wide range of biological activity of *Solidago virgaurea.* The best-known therapeutic effects of this plant are the diuretic and purifying ones. The active compounds contained in this plant, mainly flavonoids and saponins, increase the filtration in renal corpuscles and reduce the reverse resorption in renal tubules [3,8]. The astringent and anti-inflammatory properties of *Solidago virgaurea,* originating mainly from the tannins and polyphenolic acids, are used in the treatment of alimentary tract ailments [5,9]. The literature data also indicate the attractive cosmetic properties of *Solidago virgaurea,* first of all, its anti-inflammatory effect and the strengthening of blood vessels in the skin. The content of tannins, polyphenolic acids and flavonoids [3,10,11] suggests that *Solidago virgaurea* can be used for the improvement of blood circulation in the skin and its coloring [12]. Moreover, the high content of tannins and saponins in this plant implies the antiseptic properties of its extract applied on the skin and its wound healing effect [7,13]. The content of antioxidants, rutoside and quercetin from the group of flavonoids even up to 1.5 wt.% of dry product, indicates the potential anti-free radical effects of this plant, which are related to a range of anti-aging effects on the skin when applied in cosmetics [10–12]. Another interesting plant extract (from Fabaceae) with beneficial properties in the context of the skin is *Medicago sativa* extract. It owes its effects mainly to its content of a number of vitamins (A, D, E, K and B group) as well as carotenoids, flavonoids and phenolic compounds [14,15]. The high content of the mentioned biologically active substances indicates the effective antioxidant activity of *Medicago sativa* extract, acting as a free radical inhibitor and having the ability to reduce intracellular oxidative stress. What is more, it has been shown to increase skin cell proliferation and metabolism in vitro [16].

The aim of the study presented In this paper was to evaluate the potential of *Solidago virgaurea* extract as a natural cosmetic raw product applied in the form of cosmetic macroemulsions. The examination was performed taking into account technological and application aspects.

2. Materials and Methods

2.1. *Solidago Virgaurea Extract*

The extract was bought from Naturex S.c. (Katowice, Poland). The total phenolic compound (TPC) was determined by the Folin–Ciocalteu method [17] and is equal to 57.24 ± 1.49 mg/100 mL. The total flavonoids content (TFC) was assessed by the aluminum chloride colorimetric method [18], and it is equal to 57.00 ± 1.56 mg/100 mL. Although the extract contains a lower amount of polyphenols and flavonoids than described in the literature, the glycolic extract is preferred due to higher miscibility with water and thus easier formation of macroemulsions.

2.2. *Cosmetic Macroemulsions*

This study was performed on two cosmetic formulations produced by Pollena-Ewa S.A. (Zelów, Poland), labeled as macroemulsions 1 and 2: (i) day face cream SPF 30 and (ii) energizing serum for the body, improving the firmness of the skin. Each of them contained the *Solidago virgaurea* extract in the amount of 1.0 wt.% (concentration of active substances most commonly used in cosmetic products). The analogous cosmetic preparations without the extract had earlier been thoroughly examined and classified as stable emulsions. The INCI compositions of the macroemulsions investigated are shown below. The ingredients were grouped according to their content in the cosmetic preparation and assigned to one of the following ranges: (A) 3–10 wt.%; (B) 1–3 wt.%; (C) 0.3–1 wt.%; (D) 0.1–0.3 wt.%; (E) <0.1 wt.%; and (F) traces.

2.2.1. Day Face Cream SPF 30

Aqua (add to 100 wt.%), Ethylhexyl Methoxycinnamate (A), Octocrylene (A), Butyl Methoxydibenzoylmethane (A), Ethylhexyl Palmitate/Ethylhexyl Stearate/Hydrogenated Olive Oil Unsaponifiables/Caprylic/Capric Triglyceride (A), Glycerin/Brassica Napus Extract (B), C12–15 Alkyl Benzoate (B), Glycerin (B), Dicaprylyl Carbonate (B), Hydrogenated Polydecene (B), Canola Oil (B), Sorbitol/Dihydroxymethylchromone (B), Cetearyl Alcohol (B), Diethylamino Hydroxybenzoyl Hexyl Benzoate (B), Glyceryl Stearate (B), Isopropyl Isostearate (B), Isostearyl Isostearate (B), Aluminum Starch Octenylsuccinate (B), Disodium Cetearyl Sulfosuccinate (B), Camelina Sativa Seed Oil (C), Propylene Glycol/Solidago Virgaurea Extract (C), Propylene Glycol/Medicago Sativa Extract (C), Niacinamide (C), Pentaerythrityl Distearate (C), Phenylbenzimidazole Sulfonic Acid (C), Silica/CI 77891 (Titanium Dioxide)/CI 77491 (Iron Oxides) (C), Phenoxyethanol/Ethylhexylglycerin (C), Triethanolamine (C), Butyrospermum Parkii (Shea) Butter (C), Glyceryl Stearate Citrate (C), Sodium Polyacrylate (C), Glycerin/Artemia Extract (D), Tocopheryl Acetate (D), Parfum (D), Benzyl Alcohol (D), Phenylpropanol (D), Xanthan Gum (D), Tetrasodium EDTA (E) and 2-Bromo-2-Nitropropane-1,3-Diol (E).

2.2.2. Energizing Serum for the Body

Aqua (add to 100 wt.%), Glycerin (A), Isopropyl Isostearate (A), Sorbitan Stearate/Sorbityl Laurate (B), Hydrogenated Polydecene (B), Camelina Sativa Seed Oil (B), Glycerin/Brassica Napus Extract (B), Propylene Glycol/Medicago Sativa Extract (C), Propylene Glycol/Solidago Virgaurea Extract (C), Canola Oil (C), Myristyl Myristate (C), Acrylates/Beheneth-25 Methacrylate Copolymer (C), Phenoxyethanol/Ethylhexylglycerin (C), Dimethicone (C), Brassica Alba Sprout Extract/Capsaicin/Caprylic/Capric Triglyceride (C), Calophyllum Tacamahaca Seed Oil (C), Propanediol/Glycerin/Agastache Mexicana Flower/Leaf/Stem Extract (D), Parfum (D), Xanthan Gum (D), Tetrasodium EDTA (E), Sodium Hydroxide (E), 2-Bromo-2-Nitropropane-1,3-Diol (E), Benzyl Alcohol (D) and Phenylpropanol (D).

2.3. *Physicochemical Characterization*

2.3.1. pH

The pH values of the macroemulsions were measured using an EcoSense® pH 10 pH/Temperature Meter, Pen Style (VMR International, Radnor, OH, USA), on day 0 and day 30, for samples stored at room temperature. A measuring electrode was immersed in the macroemulsion, and after about 10 s, the indication was read off. The measurements were repeated 3 times for each formulation; the accepted result was the arithmetic mean and the standard deviation.

2.3.2. Viscosity

The viscosity of the macroemulsions was measured by a rotational viscosimeter RC02 (Rheotec, Dresden, Germany), on day 0 and day 30, for samples stored at room temperature. The measurements were performed using subsequent spindles inserted into the sample and rotated with a speed changed until obtaining the indication of 75% on the control unit. For each macroemulsion, the measurement was repeated three times to calculate the arithmetic mean and standard deviation.

2.4. *Particle Size Distribution*

The particle size distributions were measured for the two macroemulsions with the use of a Mastersizer 2000 (Malvern Panalytical, Columbia, MD, USA) analyzer. The measurements were conducted on day 0 and day 30 for samples stored at room temperature. A portion of 0.1 g of a cosmetic formulation was dispersed in 800 mL of distilled water at a stirring rate of 1500 rpm, maintaining the obscuration values set at the assumed level of 5%.

For each sample, the measurement was repeated three times. The results are displayed as the particle size distribution curves. Also, five values of the d(0.9) parameter were obtained and used for the determination of the arithmetic mean and standard deviation. It should be mentioned that d(0.9) is the intercept for 90% of the cumulative mass and describes the diameter where 90% of the distribution has a smaller particle size and 10% has a larger particle size.

2.5. Stability Test

The stability of the macroemulsions studied was evaluated by the method of multiple light scattering using a Turbiscan Lab Expert analyzer (Formulation, L'Union, France). Measurements were conducted on day 0 for samples stored at three selected temperatures of 4, 25 and 40 °C and then repeated at 7-day intervals, on the 7th, 15th and 30th day. The sample studied was placed in a glass vial taking care to keep a flat meniscus. The sample in the vial was scanned by a beam of infrared radiation generated by the head of the analyzer. For each sample, the measurements were repeated four times. The results were obtained as curves illustrating changes in the light backscattering and the Turbiscan Stability Index (TSI) calculated by the device software.

2.6. Skin Condition Evaluation

The effectiveness of the macroemulsions studied was evaluated on a group of 20 female volunteers of 18 to 30 years of age meeting the criteria of choice of participants. The testing lasted for four weeks. The climatic conditions in the study room, that is, the air temperature and humidity, were maintained in the recommended ranges. The in vivo measurements were approved by the Bioethical Commission no. 140/17 of 1 February 2018. The volunteers were asked to test two cosmetic macroemulsions labeled as 1 face cream and 2 serum for the body. The macroemulsions were applied according to the assumed rules of application. Prior to the experimental studies, the volunteers expressed formal consent to take part in this study and obtained detailed information on the conditions of the experimental observations (aim, rules of application, range and recommendations).

2.6.1. Skin Moisturization Measurement

The level of epidermis moisturization was measured using a Corneometer® CM 825 (Courage+Khazaka electronic GmbH, Cologne, Germany). The probe of the device was vertically pressed onto the epidermis at the site of the cosmetic product application; the result appeared on the monitor screen. It was obtained in the program Multiprobe, which is compatible with the Courage+Khazaka instruments. Measurements were conducted prior to the first application of the cosmetic product (day 0) and then at one-week intervals for four weeks, so in weeks 1, 2, 3 and 4. At each site on the epidermis, the measurements were repeated ten times. The results expressed in relative units [CM] were used to calculate the arithmetic mean and standard deviation.

2.6.2. Transepidermal Water Loss Measurement

The transepidermal water loss was measured with the use of a Tewameter® TM 300 (Courage+Khazaka electronic GmbH). The probe of the device was horizontally pressed onto the epidermis at the site of the cosmetic product application. At each site, 20 single measurements were conducted that were later averaged and used for statistical testing. The first data were collected prior to the first application (day 0) and then in one-week intervals, so in weeks 1, 2, 3 and 4. In each area of the skin, the measuring procedure was performed once; a single procedure included 20 measurements. The results expressed as the volume of steam secreted by the skin [$g/h \cdot m^2$] were used to obtain the arithmetic mean and standard deviation.

2.6.3. Skin Topography Measurement

Skin topography measurements were carried out using a Visioscan® VC 98 (Courage+Khazaka electronic GmbH). The probe of the instrument was vertically pressed onto the epidermis at the site of the macroemulsion application in order to obtain a good-quality image of the epidermis. On the basis of this image, the following skin topography parameters were determined: (i) SELS parameters (Ser roughness, Sesm smoothness, Sesc scaliness) and (ii) texture parameters (NRJ energy, HOM homogeneity). The image was analyzed prior to the first application (day 0) and in the 4th week of the application in vivo. At each site measured, the procedure was repeated three times, and the arithmetic mean and standard deviation were calculated.

2.6.4. Skin Macrorelief Measurement

The skin macrorelief parameters were determined with the help of a Visioline® VL 650 analyzer (Courage+Khazaka electronic GmbH). Analysis of the skin macrorelief was performed using a high-quality silicon replica of the skin at the site of application of the macroemulsion tested. On the basis of this replica, the values of selected parameters were established: the total area of wrinkles (%) and their mean length [mm] and density [μm]. The procedure was conducted prior to the first application of the cosmetic product tested and in the last week of its application (4th week). At each site, the procedure was performed once.

2.6.5. Statistical Analysis

Statistical analysis was carried out using the Wilcoxon test for pairs of observation results. The level of significance was $p < 0.05$.

3. Results and Discussion

3.1. Evaluation of Physicochemical Properties of Cosmetic Macroemulsions Tested

3.1.1. pH

The pH values were measured on day 0 and on day 30 for the macroemulsions studied stored at room temperature for 30 days. It is assumed that the pH of the macroemulsions should be slightly acidic (pH of 5.5) or neutral (pH of 6–7.7) [19] so as not to disturb the functioning of the hydrolipid skin coat. Changes in the pH during a cosmetic's storage may indicate the growing instability of a given cosmetic formulation caused by the compromised quality of the raw products used or the earlier contamination of the formulation in the process of production [20]. The changes in the pH observed for 30 days were insignificant. For macroemulsion 1 (face cream), the pH decreased by about 2%, from 6.52 ± 0.02 (day 0) to 6.38 ± 0.01 (day 30), while for macroemulsion 2 (serum for the body), the value of the pH increased from 5.27 ± 0.02 (day 0) to 5.42 ± 0.01 (day 30). The pH values of the macroemulsions studied were concluded to be correct, and the small changes in the pH over a period of 30 days confirmed the high quality of the raw products, including the *Solidago virgaurea* extract, and the effectiveness of the technological process preventing the contamination of the products.

3.1.2. Viscosity

The rheological properties of a cosmetic product determine its consistency and the application parameters, form of dosage, ease of spreading on the skin and rate of absorption. Significant changes in the viscosity of a given cosmetic product at each stage of the technological process may indicate undesirable changes related to instability [21]. The viscosity of the macroemulsions studied was measured on day 0 and on day 30. Analysis of the results of measurements showed that the viscosity of macroemulsion 1 insignificantly decreased by about 2.5%, from 23,100 ± 346 mPas (day 0) to 22,543 ± 223 mPas (day 30), measured with spindle R6 at a rotation rate of 30 rpm. The viscosity of macroemulsion 2 (serum for the body) significantly increased over the 30-day period, from 3604 ± 116 mPas (day 0) to 5983 ± 151 mPas (day 30), measured with spindle R3 at a rotation rate of 12 rpm. Such a great change in viscosity suggested the possible instability of the sample and impelled us to perform careful verification of the results of the stability tests described below.

3.2. Particle Size Distribution Analysis by Laser Diffraction

Particle size analysis, in particular, the determined parameter d(0.9) expressed in micrometers (μm) describing the diameter of 90% of the particles, provides information on the degree of homogeneity of a cosmetic product [22,23]. Figure 1 presents the particle size distribution curves measured on day 0 and day 30 for the two macroemulsions studied, Figure 1A,B.

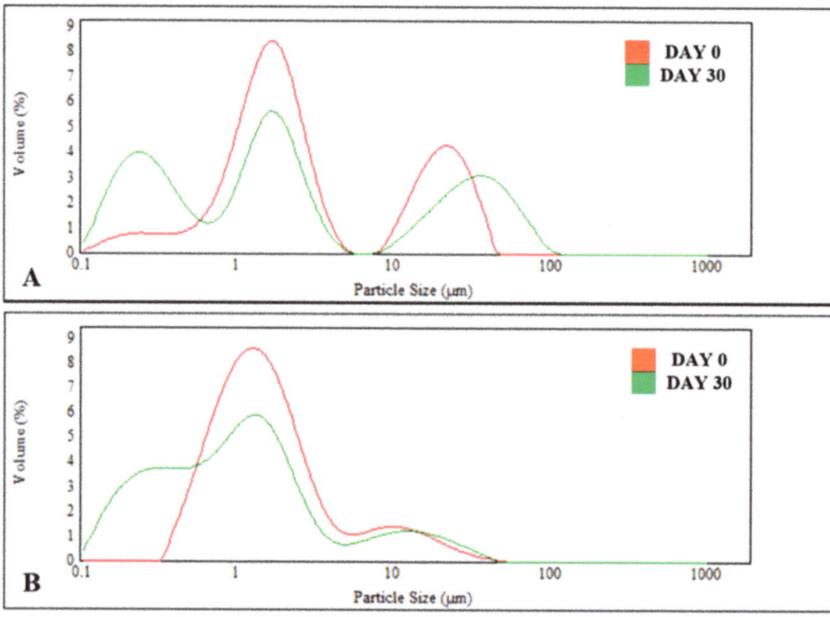

Figure 1. Comparison of the particle size distribution curves recorded for macroemulsions 1 (**A**) and 2 (**B**), stored at 25 °C for 30 days.

As shown in Figure 1, the particle size distribution curves obtained for the two samples studied shifted towards smaller particles with time, comparing the red and green curves. Moreover, for macroemulsion 1, a deterioration in homogeneity was noted, manifested as the additional maximum in the range of small particles on day 30, Figure 1A. As concerns the parameter d(0.9), for macroemulsion 1, it significantly increased over 30 days, from 30.305 ± 6.486 μm to 47.603 ± 4.045 μm, indicating a formation of agglomerations of the dispersed phase. For macroemulsion 2, the particle size distribution curve shifted with time, but its shape did not change significantly, Figure 1B. This means that the macroemulsion is stable, and this observation is confirmed by the values of the d(0.9) parameter that slightly increased from 8.248 ± 0.757 on day 0 to 8.967 ± 1.148 on day 30. Analysis of the particle size distribution changes with time for macroemulsion 1 (face cream) indicates a possible destabilization process, not visible to the naked eye, which may be a consequence of the incomplete compatibility of the *Solidago virgaurea* extract with the other components of this cosmetic product. A potential source of instability is the synthetic emulsifier used in this macroemulsion, which is an ester of benzoic acid and fatty alcohols C12–15, known to be able to undesirably decrease the viscosity of cosmetic emulsions containing it.

3.3. *Stability Study by Multiple Light Scattering*

Tests on cosmetic formulations' stability are carried out to predict their ability to maintain their original physicochemical properties over the declared shelf life [24,25], which in the case of the macroemulsions tested was set at 2 years. The use of multiple light scattering (MLS) permits the detection of changes not visible to the naked eye but changing the physical stability of a given cosmetic product [26–28]. The analysis of stability by MLS was performed for the two cosmetic macroemulsions stored at three temperatures, 4, 25 and 40 °C, for 30 days. The main parameter analyzed was the Turbiscan Stability Index (TSI), characterizing the total stability of the cosmetic product, whose value was predicted to increase with the increasing destabilization of the sample [29,30].

As follows from the results of the TSI measurements for macroemulsion 1, Figure 2A, for the sample stored at 4 °C, the TSI changes over a period of 30 days are insignificant; for the sample stored at room temperature, the TSI changes are acceptable, below a value of 3. However, for the sample stored at an elevated temperature (40 °C), the TSI value after 30 days increases almost to 4, which is high taking into account the relatively short time of storage.

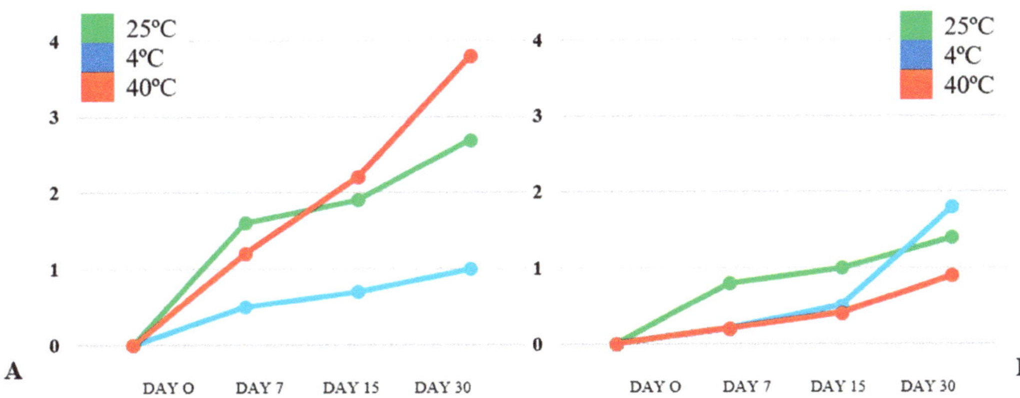

Figure 2. Changes in Turbiscan Stability Index for macroemulsions 1 (**A**) and 2 (**B**), stored at different temperatures for 30 days.

Figure 3A presents the backscattering profile (ΔBS) for macroemulsion 1 (stored at 40 °C); the x axis represents the height of the sample, while the y axis represents the intensity of backscattering. The measurements conducted on day 0 are marked in blue, while those conducted on day 30 are presented in red. The results show a decrease in the backscattering intensity, reaching up to 6%, in the lower part of the sample (left side of the profile), between the results obtained on day 0 and day 30. The change was interpreted as instability described as insignificant creaming; this observation was made only for the sample stored at an elevated temperature.

For macroemulsion 2, the TSI values were notably lower for the samples stored at all temperatures of storage (Figure 2B). These results are in full agreement with the backscattering profiles recorded for the sample stored at 40 °C (Figure 3B), which is characteristic of a cosmetic macroemulsion that is stable over time; no significant changes in the backscattering intensity were noted in the entire volume of the sample. The results of the MLS analysis correspond to the data obtained from the particle size distribution showing the relatively lower stability of macroemulsion 1, however, still at an acceptable level, which permitted the use of this cosmetic product in further in vivo tests.

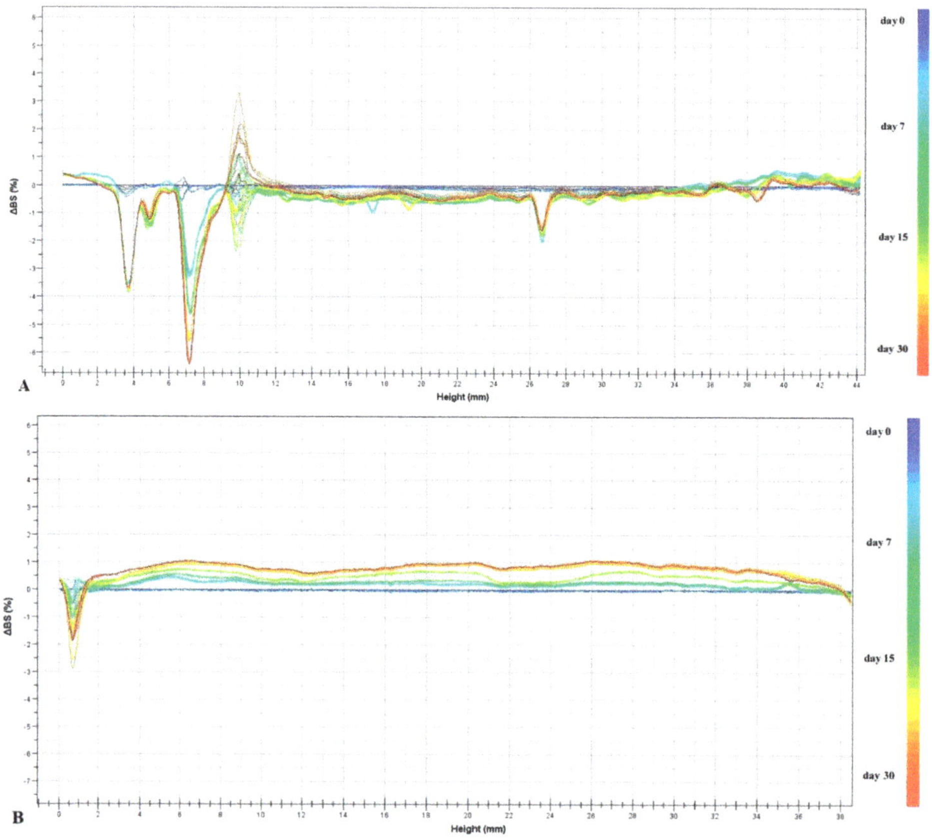

Figure 3. Delta backscattering profiles recorded for macroemulsions 1 (**A**) and 2 (**B**), stored at 40 °C for 30 days.

3.4. Skin Parameter Estimation

Prior to the in vivo examination, the effectiveness of the bases of the cosmetic products studied, without the active ingredient of *Solidago virgaurea* extract, was evaluated. As a result of this evaluation, the impact of the cosmetic bases was described as neutral.

3.4.1. Analysis of the Skin Hydration and the Transepidermal Water Loss Measurement

The degree of epidermis moisturization and the level of transepidermal water loss are the coupled parameters of the skin characterizing the ability to bind and retain water in particular skin components [31]. The true effect of skin moisturization as a result of the regular application of cosmetic care preparations can be reached only if the cosmetic product used is also able to preserve the sufficient barrier properties of the epidermis protecting against transepidermal water loss [32–34]. The expected desirable effects of cosmetic product application were described by (i) an increase in the epidermis moisturization to above 45 CM [35,36] and (ii) a reduction in the level of transepidermal water loss (TEWL), preferably below 10 $g/h{\cdot}m^2$ [37,38].

The analysis of changes in the degree of epidermis moisturization and the coupled changes in the TEWL (Table 1) should be performed in two aspects. For both macroemulsions, a beneficial decrease in the TEWL of 6.6% and 7.7% was observed for macroemulsions 1 and 2, respectively, at $p < 0.05$; however, their epidermis moisturizing effects were different. The application of macroemulsion 1 resulted in an increase in the stratum corneum moisturization of about 7% ($p < 0.05$). The effect of this cosmetic preparation on the skin expressed by the TEWL value and moisturization was described as positive; moreover, it was additionally enhanced by the strengthening of the epidermis barrier functions. The results obtained for macroemulsion 2 evidenced that its application improved the moisturization of the epidermis by over 70% ($p < 0.05$). After the 4-week application of macroemulsion 2, the skin of the volunteers at the site of application was classified as of the correct degree of moisturization (>45 CM), while at the beginning of this study, it was classified as very dry (~30 CM). This really substantial increase in the epidermis moisturization was accompanied by a significant decrease in the TEWL, which permits the conclusion that macroemulsion 2 (serum for the body) shows much stronger moisturizing properties than macroemulsion 1 (face cream). Moreover, the analysis of the results indicated that the ability to bind and retain water in the skin structures, which is the effect assigned to *Solidago virgaurea* extract, was much greater in macroemulsion 2, in which this extract was much more compatible with the other components of the cosmetic formulation, as evidenced by the earlier results of physicochemical studies and stability evaluation. Similar results were also obtained by Prof. Vamanu's research group, which, in the course of studies on extracts of *Lactarius piperatus* and *Ribes rubrum*, confirmed the beneficial effect of facial creams containing them in increasing skin hydration levels [39].

Table 1. Changes in the basic skin parameters measured during the application tests (data are expressed as mean ± SD [1] and as percentage change Δ%).

	TEWL [2] [g/h·m^2]		HYDRATION [CM [3]]	
Week	**Macroemulsion I**	**Macroemulsion II**	**Macroemulsion I**	**Macroemulsion II**
0	16.28 ± 4.67	11.48 ± 4.28	52.62 ± 8.25	30.07 ± 7.85
1	17.03 ± 1.57	12.44 ± 1.69	51.68 ± 9.53	41.73 ± 11.11
2	13.61 ± 1.20	12.38 ± 1.46	53.10 ± 7.14	48.06 ± 13.45
3	15.96 ± 1.46	10.69 ± 1.54	52.14 ± 7.28	46.14 ± 9.04
4	15.21 ± 1.56	10.60 ± 1.57	56.34 ± 5.95	51.89 ± 9.77
Δ%	−6.6	−7.7	7.1	72.6

[1] SD: standard deviation; [2] TEWL: transepidermal water loss; [3] CM: arbitrary units of skin hydration.

3.4.2. Analysis of the Skin Topography Parameters

Analysis of the skin topography permits the non-invasive and comprehensive characterization of the skin surface on the basis of the distribution of the levels of grey in a photograph of a given area of the skin surface [36,40]. The analysis provides information on the homogeneity of the epidermis and general condition of the skin related to the degree of wrinkling and moisturization of the stratum corneum [40]. The first group of skin topography parameters are the SELS ones: (i) SEr (roughness) evaluated on the basis of the dark pixels in the image, corresponding to the wrinkles and furrows in the skin; (ii) SEsm (smoothness) evaluated on the basis of the uniform character of the image, which is the degree of differences in the levels of grey in the image; and (iii) SEsc (scaliness) related to the number of white pixels in the image interpreted as the sites on the skin with a reduced moisturization degree [41,42]. The second group of skin texture parameters are energy NRJ and homogeneity HOM describing the general condition and homogeneity of the skin surface, calculated on the basis of the analysis of repetitions of color combinations in neighboring pixels and differences between the levels of grey [41–44].

Figure 4 presents the averaged changes in the skin topography parameters (expressed in %) noted between day 0 and day 30. The expected desirable direction of changes in SELS parameters was observed only for macroemulsion 2. After 4 weeks of application of this cosmetic product, the parameter SEr decreased by 17.4% ($p < 0.05$), proving the effectiveness of macroemulsion 2 as an anti-aging cosmetic, reducing visible wrinkles and furrows. The parameters SEsm and SEsc decreased by 22.1% and 10.3% ($p < 0.05$), confirming the strong moisturizing properties of this macroemulsion. As far as the other parameters are concerned, statistically significant changes were noted only in NRJ whose value increased by over 13% ($p < 0.05$), evidencing an improvement in the general skin condition. The reduction in HOM amounted to 0.6%, so was statistically insignificant. The results of the 4-week application of macroemulsion 1 (face cream) revealed its undesirable effect on the skin topography parameters. The parameters SEr and SEsm increased by 5.7% and 34.1%, respectively ($p < 0.05$), so changed in the undesired direction. The desired increase in SEsc by about 0.1% did not compensate for the above negative effect of the preparation and confirmed the much weaker moisturizing effect of macroemulsion 1 than 2. The value of NRJ showed an undesired 25% decrease ($p < 0.05$), which was consistent with the negative change in SEsm. No statistically significant change was noted in the HOM after the 4-week application test.

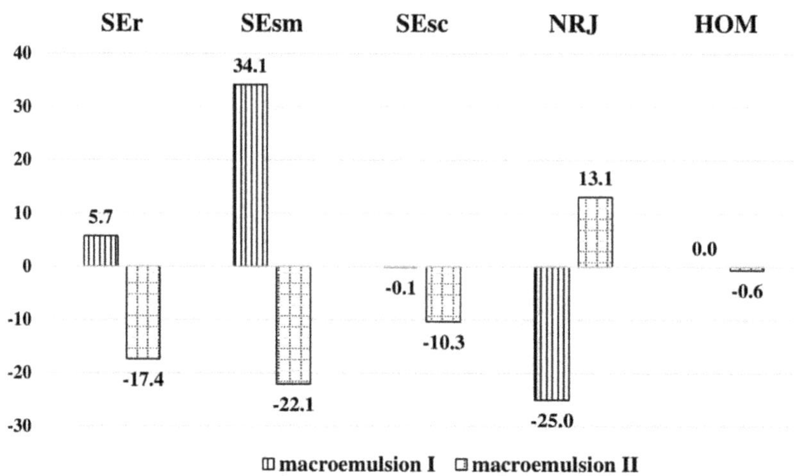

Figure 4. Percentage changes Δ% in the selected parameters of the skin topography determined in vivo as a result of the application of macroemulsions 1 and 2.

A study on the effect of cosmetic emulsions on the improvement in selected skin topography parameters was also conducted by Waqas et al.—the application properties of cosmetics containing 4 wt.% soybean extract were examined. The trend of their results was analogous to that in our study, i.e., there was a decrease in the values of selected skin topography parameters (SEr by 5%, SEsm by 6%, SEsc by 4%.). However, the cited research results were not as satisfactory as in the case of the study using *Solidago virgaurea* extract [45].

3.4.3. Analysis of the Skin Macrorelief Parameters

Analysis of the skin macrorelief parameters permits the evaluation of the skin structure and, in particular, the verification of the anti-aging effect of a given cosmetic product [41]. The desired effect was assumed to be a reduction in the values of the total wrinkle area and the wrinkles' mean length and depth, determined on the basis of a special way of lighting the skin replicas [44,46].

Analysis of the changes in the skin macrorelief parameters after the 4-week application of the macroemulsions tested, Figure 5, showed their positive effect; however, the effect of macroemulsion 2 was more positive, in which the *Solidago virgaurea* extract was more compatible with its other components. The 4-week application of macroemulsion 2 resulted in a nearly 2% ($p < 0.05$) greater reduction in the total wrinkle area and an over 12% greater reduction ($p < 0.05$) in the mean wrinkle length than that observed for macroemulsion 1. Only the mean wrinkle depth was reduced to a greater degree by macroemulsion 1 application than the application of macroemulsion 2; however, the difference was statistically insignificant (of 1%), indicating no differences in this effect between the two cosmetic products. The results of the macroemulsions' effects on the skin macrorelief are not fully consistent with those concerning the skin topography. Nevertheless, the results definitely point to the intense anti-aging and moisturizing effects of macroemulsion 2. The desirable changes in the total wrinkle area and the wrinkles' mean length are fully consistent with the reduction in the values of the skin topography parameters SEr, SEsm and SEsc.

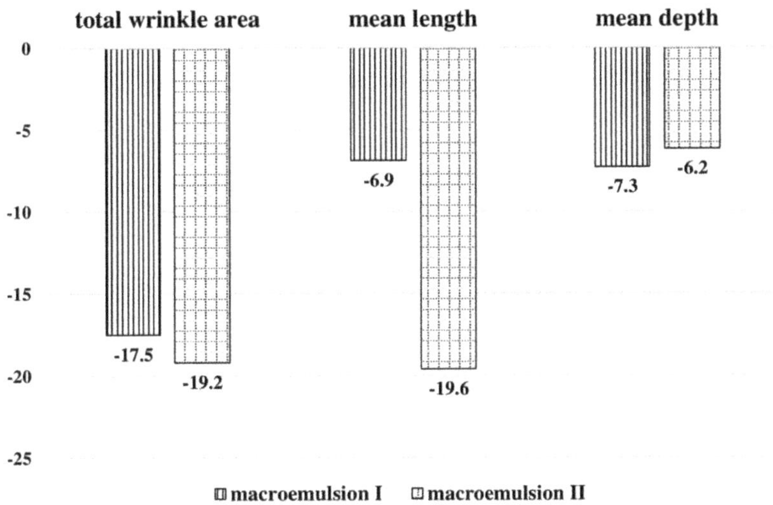

Figure 5. Percentage changes Δ% in selected parameters of the skin macrorelief determined in the in vivo tests of macroemulsions 1 and 2.

4. Conclusions

The therapeutic properties of *Solidago virgaurea* extract have encouraged us to check the suitability and effectiveness of using it as an active ingredient in cosmetic macroemulsions. The verification of the applicability of *Solidago virgaurea* extract as a cosmetic active ingredient was a challenge because of the complexity of physicochemical and in vivo studies required prior to the introduction of cosmetic products containing this extract to the market. The results of the comprehensive studies permitted the evaluation of the physicochemical and application properties of two macroemulsions containing *Solidago virgaurea* extract and indicated a correlation between the presence of *Solidago virgaurea* extract in the cosmetic product and its positive effect on selected skin parameters. The introduction of the extract had no effect on the physicochemical parameters and stability of the macroemulsion in which the extract was highly compatible with its other components, so of macroemulsion 2 applied in the form of a serum for the body. The compatibility of *Solidago virgaurea* extract with the base of the cosmetic product was found to significantly affect the effectiveness of the cosmetic product. The final rheology of the cosmetic microemulsion had a significant impact on the application possibilities of the cosmetic product. The determining factor was the choice of the emulsifier. In microemulsion 2, the emulsifier was Sorbitan Stearate, and the effectiveness of this product was much higher than that of the cosmetic formulations containing other emulsifiers (synthetic ester of benzoic acid or fatty alcohols C12–15). The results of this experimental study have shown the importance of the right choice of base for cosmetic formulations and its compatibility with the natural plant extracts to be introduced for the products' effectiveness. It is also important to mention the necessity to extend the discussed studies in the future, i.e., to perform long-term stability studies of finished cosmetic macroemulsions containing *Solidago virgaurea* extract, and to consider conducting in vivo tests on a group of volunteers with sensitive skin in order to confirm the versatility of the obtained cosmetic emulsions that do not cause adverse allergic reactions.

Author Contributions: Conceptualization, M.M. and I.N.; methodology, M.M.; validation, M.M.; investigation, M.M.; resources, G.K., W.S. and I.N.; writing—original draft preparation, M.M.; writing—review and editing, M.M. and I.N.; visualization, M.M.; supervision, I.N.; project administration, I.N.; funding acquisition, I.N. All authors have read and agreed to the published version of the manuscript.

Funding: This work was supported by the National Centre for Research and Development, grant number no. BIOSTRATEG2/298205/9/NCBR/2016, "Cultivated plants and natural products as a source of biologically active substances assigned to the production of cosmetic and pharmaceutical products as well as diet supplements".

Institutional Review Board Statement: This study was conducted according to the guidelines of the Declaration of Helsinki and approved by the Bioethical Commission of Poznan University of Medical Sciences (140/17, 1 February 2018).

Informed Consent Statement: Informed consent was obtained from all subjects involved in this study.

Data Availability Statement: The raw data supporting the conclusions of this article will be made available by the authors on request. The general procedure of the macroemulsions' formation is only available to non-professionals.

Acknowledgments: The authors are grateful to Paulina Radoszewska and Agnieszka Sobczak—students working towards a master's degree—for their help in the experiment.

Conflicts of Interest: Authors Grażyna Kaszczyk and Witold Sujka were employed by TRICOMED SA. The remaining authors declare that this research was conducted in the absence of any commercial or financial relationships that could be construed as potential conflicts of interest.

References

1. Karpaviciene, B.; Radušienė, J. Morphological and Anatomical Characterization of *Solidago* × *niederederi* and Other Sympatric Solidago Species. *Weed Sci.* **2016**, *64*, 61–70. [CrossRef]
2. Kopeć, D.; Michalska-Hejduk, D.; Polak, A.; Polak, P. Gatunki z rodzaju nawłoć *Solidago* spp. In *Inwazyjne Gatunki Roślin w Kampinoskim Parku Narodowym i Jego Sąsiedztwie*; Michalska-Hejduk, D., Otręba, A., Eds.; Kampinoski Park Narodowy: Izabelin, Poland, 2016; Chapter 5.2.2.
3. Moricz, A.M.; Ott, P.G.; Habe, T.T.; Darcsi, A.; Boszormenyi, A.; Alberti, A.; Kruzselyi, D. Effect-directed discovery of bioactive compounds followed by highly targeted characterization, isolation and identification, exemplarily shown for *Solidago virgaurea*. *Anal. Chem.* **2016**, *88*, 8202–8209. [CrossRef] [PubMed]
4. Moslavac, T.; Jokić, S.; Subarić, D.; Aladić, K.; Vukoja, J.; Prce, N. Pressing and supercritical CO_2 extraction of *Camelina sativa* oil. *Ind. Crops Prod.* **2014**, *54*, 122–129. [CrossRef]
5. Woźniak, D.; Ślisarczyk, S.; Domaradzki, K.; Dryś, A.; Matkowski, A. Comparison of Polyphenol Profile and Antimutagenic and Antioxidant Activities in Two Species Used as Source of *Solidaginis herba*—Goldenrod. *Chem. Biodivers.* **2018**, *15*, e1800023. [CrossRef]
6. Sabir, S.M.; Ahmad, S.D.; Hamid, A.; Khan, M.Q.; Athayde, M.L. Antioxidant and hepatoprotective activity of ethanolic extract of leaves of *Solidago microglossa* containing polyphenolic compounds. *Food Chem.* **2012**, *131*, 741–747. [CrossRef]
7. Starks, C.M.; Williams, R.B.; Goering, M.G.; O'Neil-Johnson, M.; Norman, V.L.; Hu, J.F.; Garo, E.; Hough, G.W.; Rice, S.M.; Eldridge, G.R. Antibacterial clerodane diterpenes from Goldenrod (*Solidago virgaurea*). *Phytochemistry* **2010**, *71*, 104–109. [CrossRef]
8. Kalemba, D. Constituents of the essential oil of *Solidago virgaurea* L. *Flavour. Fragr. J.* **1998**, *13*, 373–376. [CrossRef]
9. Ahmad, A.; Abuzinadah, M.F.; Alkreathy, H.M.; Kutbi, H.I.; Shaik, N.A.; Ahmad, V.; Saleem, S.; Husain, A. A novel polyherbal formulation containing thymoquinone attenuates carbon tetrachloride-induced hepatorenal injury in a rat model. *Asian Pac. J. Trop. Biomed.* **2020**, *10*, 147–155. [CrossRef]
10. Demir, H.; Acik, L.; Bali, E.B.; Koc, L.Y.; Kaynak, G. Antioxidant and antimicrobial activities of *Solidago virgaurea* extracts. *Afr. J. Biotechnol.* **2009**, *8*, 274–279.
11. Ji, Y.S.; Lee, N.S.; Kil, K.J.; Yoo, J.H. Antioxidant Activities of Ethanol Extracts from *Solidago virgaaurea* var. gigantea. *Korea J. Herbol.* **2019**, *34*, 109–116. [CrossRef]
12. Mietlińska, K.; Przybyt, M.; Kalemba, D. Polish plants as raw materials for cosmetic purposes. *Biotechnol. Food Sci.* **2019**, *83*, 95–107. [CrossRef]
13. Kołodziej, B.; Kowalski, R.; Kędzia, B. Antibacterial and antimutagenic activity of extracts aboveground parts of three Solidago species: *Solidago virgaurea* L., *Solidago canadensis* L. and *Solidago gigantea* Ait. *J. Med. Plant Res.* **2011**, *5*, 6770–6779. [CrossRef]

14. Bora, K.S.; Sharma, A. Phytochemical and pharmacological potential of *Medicago sativa*: A review. *Pharm. Biol.* **2011**, *49*, 211–220. [CrossRef] [PubMed]
15. Caunii, A.; Pribac, G.; Grozea, I.; Gaitin, D.; Samfira, I. Design of optimal solvent for extraction of bio-active ingredients from six varieties of *Medicago sativa*. *Chem. Cent. J.* **2012**, *6*, 123. [CrossRef] [PubMed]
16. Zagórska-Dziok, M.; Ziemlewska, A.; Nizioł-Łukaszewska, Z.; Bujak, T. Antioxidant activity and cytotoxicity of *Medicago sativa* L. seeds and herb extract on skin cells. *BioRes.h Open Access* **2020**, *9*, 229–242. [CrossRef]
17. Kaur, C.; Kapoor, H.C. Anti-oxidant activity and total phenolic content of some Asian vegetables. *Int. J. Food Sci. Technol.* **2002**, *37*, 153–161. [CrossRef]
18. Arvouet-Grand, A.; Vennat, B.; Pourrat, A.; Legret, P. Standardization of propolis extract and identification of principal constituents. *J. Pharm. Belg.* **1994**, *49*, 462–468.
19. Lambers, H.; Piessens, S.; Bloem, A.; Pronk, H.; Finkel, P. Natural skin surface pH is on average below 5, which is beneficial for its resident flora. *Int. J. Cosmet. Sci.* **2006**, *28*, 359–370. [CrossRef]
20. Schmid-Wendtner, M.H.; Korting, H.C. The pH of the skin surface and its impact on the barrier function. *Skin Pharmacol. Physiol.* **2006**, *19*, 296–302. [CrossRef]
21. Dąbrowska, M. Optimization of Physicochemical and Application Properties of Cosmetic Formulations Containing Selected Iridoid Glycosides. Ph.D. Thesis, Faculty of Chemistry, Adam Mickiewicz University,, Poznań, Poland, 2019.
22. Eshel, G.; Levy, G.J.; Mingelgrin, U.; Singer, M.J. Critical Evaluation of the Use of Laser Diffraction for Particle-Size Distribution Analysis. *Soil Sci. Soc. Am. J.* **2004**, *68*, 736–743. [CrossRef]
23. MD Malvern Instruments. *Rapid Refractive Index Determination for Pharmaceutical Actives*; Application Note; Malvern Panalytical Ltd.: Columbia, MD, USA, 2004.
24. *Regulation (EC) No 1223/2009 of the European Parliament and of the Council of 30 November 2009 on Cosmetic Products*; European Parliament: Strasbourg, France, 2009.
25. Gościańska, J.; Olejnik, A.; Nowak, I. *Analityka Środków Kosmetycznych*; Cursiva: Kostrzyn, Poland, 2013.
26. Madaan, V.; Chanana, A.; Kataria, M.K.; Bilandi, A. Emulsion technology and recent trends in emulsion applications. *Int. Res. J. Pharm.* **2014**, *5*, 533–542. [CrossRef]
27. Salager, J.; Bullón, J.M.; Pizzino, A.; Rondón-González, M.; Tolosa, L.I. Emulsion Formulation Engineering for the Practitioner. In *Surface and Colloid Science*; Matijevic, E., Borkovec, M., Eds.; Taylor & Francis: New York, NY, USA, 2010.
28. Morrison, I. Emulsion Technology. In Proceedings of the ACS National Meeting, New Orleans, LA, USA, 9–10 April 2008.
29. Bru, P.; Brunel, L.; Buron, H.; Cayré, I.; Ducarre, X.; Fraux, A.; Mengual, O.; Meunier, G.; de Sainte Marie, A.; Snabre, P. Particle Size and Rapid Stability Analyses of Concentrated Dispersions: Use of Multiple Light Scattering Technique. In *Particle Sizing and Characterization*; Provder, T., Texter, J., Eds.; American Chemical Society: Washington, DC, USA, 2004; pp. 45–60.
30. Formulaction. *User Guide Turbiscan Lab Expert*; Microtrac Inc.: York, PA, USA, 2013.
31. Olejnik, A.; Schroeder, G.; Nowak, I. The tetrapeptide N-acetyl-Pro-Pro-Tyr-Leu in skin care formulations—Physicochemical and release studies. *Int. J. Pharm.* **2015**, *492*, 161–168. [CrossRef] [PubMed]
32. Verdier-Sévrain, S.; Bonté, F. Skin hydration: A review on its molecular mechanisms. *J. Cosmet. Dermatol.* **2007**, *6*, 75–82. [CrossRef] [PubMed]
33. Roberts, M.S.; Cross, S.E.; Pellett, M.A. Skin Transport. In *Dermatological and Transdermal Formulations*; Walters, K.A., Ed.; Marcel Dekker: New York, NY, USA, 2002; pp. 89–115.
34. Kmieć, M.L.; Pajor, A.; Broniarczyk-Dyła, G. Evaluation of biophysical skin parameters and assessment of hair growth in patients with acne treated with isotretinoin. *Adv. Dermatol. Allergol.* **2013**, *30*, 343–349. [CrossRef]
35. Baumann, L. Understanding and treating various skin types: The Baumann Skin Type Indicator. *Dermatol. Clin.* **2008**, *26*, 359–373. [CrossRef]
36. Barel, A.O.; Paye, M.; Maibach, H.I. (Eds.) *Handbook of Cosmetic Science and Technology*, 3rd ed.; Informa Healthcare: New York, NY, USA, 2009.
37. Fluhr, J.W.; Gloor, M.; Lazzerini, S.; Kleesz, P.; Grieskaber, R.; Berardesca, E. Comparative study of five instruments measuring stratum corneum hydration. *Skin. Res. Technol.* **1999**, *5*, 161–170. [CrossRef]
38. Gardien, K.L.M.; Baas, D.C.; De Vet, H.C.W.; Middelkoop, E. Transepidermal water loss measured with the Tewameter TM300 in burn scars. *Burns* **2016**, *42*, 1455–1462. [CrossRef]
39. Papa, C.M.; Suciu, A.; Dopcea, I.; Ene, N.; Singh, S.K.; Vamanu, E. Exploring the Efficacy of Extracts for Cosmetic Creams: In Vivo and In Vitro Assessments. *Nutraceuticals* **2023**, *3*, 306–314. [CrossRef]
40. Dąbrowska, M.; Mielcarek, A.; Nowak, I. Evaluation of sex-related changes in skin topography and structure using innovative skin testing equipment. *Skin Res. Technol.* **2018**, *24*, 614–620. [CrossRef]
41. Trojahn, C.; Dobos, G.; Schario, M.; Ludriksone, L.; Blume-Peytavi, U.; Kottner, J. Relation between skin micro-topography, roughness, and skin age. *Skin Res. Technol.* **2015**, *21*, 69–75. [CrossRef]
42. Tronnier, H.; Heinrich, U.; Stute, R.; Mallia, C.; Uitto, J. Surface evaluation of Living Skin. In *Rheumaderm*; Mallia, C., Uitto, J., Eds.; Kluwer Academic, Plenum Publishers: New York, NY, USA, 1999; pp. 507–516.
43. Dzwigałowska, A.; Sołyga-Żurek, A.; Dębowska, R.M.; Eris, I. Preliminary study in the evaluation of anti-aging cosmetic treatment using two complementary methods for assessing skin surface. *Skin Res. Technol.* **2013**, *19*, 155–161. [CrossRef] [PubMed]

44. Houser, T.; Zerweck, C.; Grove, G.; Wickett, R. Shadow analysis via the C+K Visioline: A technical note. *Skin Res. Technol.* **2017**, *23*, 447–451. [CrossRef] [PubMed]
45. Waqas, M.K.; Akhtar, N.; Rasul, A.; Rashid, S.U.; Mustafa, R.; Khan, B.A.; Murtaza, G. In vivo Evaluation of a Cosmetic Emulsion Containing Soybean Extract for Anti-Aging. *Trop. J. Pharm. Res.* **2014**, *13*, 1401–1406. [CrossRef]
46. Quatresooz, P.; Thirion, L.; Piérard-Franchimont, C.; Piérard, G.E. The riddle of genuine skin microrelief and wrinkles. *Int. J. Cosmet. Sci.* **2006**, *28*, 389–395. [CrossRef]

applied sciences

MDPI

Article

Plant-Origin Additives from *Boswellia* Species in Emulgel Formulation for Radiotherapy Skin Care

Agnieszka Kulawik-Pióro [1,*], Weronika Goździcka [1,2], Joanna Kruk [3] and Anna Piotrowska [4]

1 Department of Organic Chemistry and Technology, Faculty of Chemical Engineering and Technology, Cracow University of Technology, Warszawska 24, 31-155 Cracow, Poland; weronikagozdzicka@windowslive.com
2 Management of Technology, Technische Universiteit Delft, Postbus 5, 2600 AA Delft, The Netherlands
3 Department of Engineering and Machinery for Food Industry, Faculty of Food Technology, University of Agriculture in Krakow, Balicka 122, 30-149 Krakow, Poland; joanna.kruk@urk.edu.pl
4 Department of Chemistry and Biochemistry, Institute of Basic Sciences, Faculty of Rehabilitation, University of Physical Education in Krakow, Jana Pawła II 78, 31-571 Cracow, Poland; anna.piotrowska@awf.krakow.pl
* Correspondence: agnieszka.kulawik-pioro@pk.edu.pl; Tel.: +48-12-628-27-59

Abstract: The research objective of this study was to include plant-origin additives of the *Boswellia* species in the formulation of topical preparations for skin care after radiotherapy. The main factor damaging the skin during radiotherapy is the free radicals that form from water molecules and granulocytes in the inflammatory area; hence, the use of substances with antioxidant properties, including plant extracts rich in antioxidants, seems to be an alternative therapy in radiodermatitis treatment. A series of cosmetic preparations containing plant-origin additives from *Boswellia* species and corresponding placebo formulations were prepared. In order to assess the applicability of emulgels as oncocosmetics, their stability, physicochemical properties, rheological properties, and antioxidant capacity were determined. Somatosensory analysis was also performed. An attempt was also made to correlate the effect of plant-derived additives on the functional properties of the product determined via instrumental methods and the sensory properties. The most promising preparation was the emulgel containing the Soxhlet extract and essential oil (Em_SO) due to its high antioxidant properties compared to other preparations (% inhibition of 11.69) and polyphenol content (3.63 mg/dm^3). Additionally, probands positively assessed all its features, including consistency (4.00), absorption (4.43), and hydration (4.71). The presence of significant correlations for % inhibition and polyphenols content with sensory and physicochemical characteristics of samples was indicated. There were low ratings for placebo preparations by probands, and the demonstrated correlations of odor with moisturization and distribution, oiliness and hydration with the % inhibition of the sample, and the content of polyphenols with the pH and size of the dispersed phase droplets proved the positive effect of the addition of plant-origin additives from *Boswellia* to the emulgel formulation on the functional and sensory properties.

Keywords: emulgels; radiotherapy skin care; *Boswellia* species; antioxidant properties

1. Introduction

Radiotherapy is one of the basic methods used in cancer treatment of the prostate, breast, lung, cervix, and head, which uses ionizing radiation to destroy cancer cells. With this type of therapy, it is possible to prevent recurrence, slow down, or completely stop tumor growth; hence, radiotherapy is currently included in most comprehensive cancer control plans [1–3]. According to the World Health Organization (WHO), over 50% of oncology patients require radiotherapy as part of their treatment. It is therefore used together with surgery, both before and after the surgery, and with systemic chemotherapy [3]. The widespread use of radiotherapy in oncological treatment is connected with its high efficiency [4]. Unfortunately, patients undergoing radiotherapy suffer from a number of side effects that lower their quality of life and impede their daily functioning.

Among the most common types of side effects of therapy affecting approximately 95% of patients are skin reactions—radiodermatitis [2,5]. These skin manifestations range from slight to severe skin damage. The most vulnerable areas are those parts of the body where two skin surfaces meet (breasts, crotch), those with thin and smooth skin (crotch, face, armpits), and areas of skin where the integrity of the skin layers has already been compromised [2]. The skin reactions observed during treatment depend both on the radiation dose and duration of treatment but also on comorbidities, age, general health, addictions, and other drugs used [2,6]. The most common skin reactions include transient and persistent erythema, moist and dry desquamation accompanied by serous effusion, persistent epilation, hyperpigmentation, ulceration, spots and/or rash, skin atrophy, and others [2,7–9]. Radiodermatitis also causes burning, pain, and discomfort among patients [2,8,9]. This decreases the quality of patients' lives, causing pain, treatment delay, or aesthetic defects. Radiotherapy causes damage to Langerhans cells, basal cells, and the vascular endothelium. The decreased number of Langerhans cells and the depletion of basal layer stem cells lead to impaired barrier and immune function, increasing the risk of wound infection. Damage to vasculature can induce hypoxia and TGF-β production, further driving fibrosis. Tissue hypoxia with associated necrosis and inflammation can lead to the generation of significant amounts of reactive oxygen species (ROS). Additionally, the ROS can drive the production of cytokines, perpetuating the cycle of inflammatory changes [10,11]. Radiotherapy breaks down the body's immune system and leads to an excess of ROS in the body, which is why patients use alternative methods of treatment that include antioxidants [10]. Therapies targeting the elimination of ROS include both systemic and topical delivery of antioxidants, but systemically derived antioxidants could disrupt radiation treatment therapy by preventing the formation of ROS within the tumor. The topical application of antioxidants allows for the concentrated therapy to the skin, minimizing systemic levels [10].

The ability of skin to regenerate decreases significantly when the patient's treatment lengthens or when the received radiation dose increases. The patient's skin after radiation therefore requires a special course of treatment due to its sensitivity and susceptibility to external factors such as UV radiation [2,12].

There are many skin care products on the market for use on the skin after radiotherapy that are commonly called oncocosmetics or oncology skin care products. They can be divided according to their mode of action. They include moisturizing, emollient, anti-inflammatory, cleansing, and antiseptic agents but also drying agents and barrier measures. The preparations should also stimulate the proliferation of fibroblasts and keratinocytes in order to accelerate cell renewal and, consequently, skin regeneration, and they should not include potentially allergenic ingredients [2,12–14]. The products are available on the market in such physicochemical forms as aqueous solutions, foams, gels, emulsions, suspensions, oils, pastes, and others. They have a number of advantages, including the relief of side effects such as itching, burning, redness, and retention of moisture in the skin. By improving the barrier functions of the skin, they reduce potential friction and mechanical skin damage. Disadvantages include being too oily to feel, difficulty in spreading on painful skin, comedogenic action, skin pH disruption causing irritation and unpleasant odor, intensifying nausea, and vomiting [2,12,15–17].

The most common active ingredients in the formulation of oncocosmetics include ceramides, phospholipids, hyaluronic acid, urea, D-panthenol, allantoin, vegetable and animal oils, lipids, proteins, vitamins, and essential unsaturated fatty acids, but also plant and herbal extracts and essential oils [2,18]. The beneficial effect of plant-origin additives is due to the presence in their composition of phytochemicals such as saponins, flavonoids, tannins, alkaloids, and terpenoids. These compounds are known for their anti-inflammatory, antibacterial, and anticancer properties [19]. Due to these properties, many plants and herbs can support the natural repair mechanisms of the skin and exhibit great therapeutic potential in wound care [20].

The main factor damaging the skin during radiotherapy is the free radicals that form from water molecules and granulocytes in the inflamed area; hence, the use of substances with antioxidant properties, including plant and herbal extracts rich in antioxidant substances, seems to be a complementary therapy in radiodermatitis treatment [2,17,21,22]. Plant and herbal extracts introduced to the formulations of topical medicinal and cosmetic preparations show antioxidant effects, reducing irritation and redness. They can also promote healing processes or protect against UV radiation, improving the quality of life of patients suffering from radiodermatitis [23].

A previously conducted literature analysis of the use of plant and herbal extracts as ingredients in topical formulations in the prevention and treatment of radiodermatitis [2] showed the promising effects of numerous plant and herbal extracts on skin affected by radiodermatitis, particularly *Boswellia* oleoresin, which has antioxidant, anti-inflammatory, and antimicrobial properties. The research objective of this study was therefore to include plant-origin additives of the *Boswellia* species in the formulation of topical preparations for skin care after radiotherapy. A representative belonging to the polymeric gels, i.e., emulgel, was chosen as the physicochemical form of the preparation. In order to assess the applicability of emulgels as oncocosmetics, their stability, physicochemical properties, rheological properties, and antioxidant capacity were determined. Sensory analysis was also performed.

2. Materials and Methods

The effect of the addition of plant-origin additives from *Boswellia* species on the physicochemical properties of the formulation for topical use—while maintaining antioxidant properties, which are a measure of the effectiveness of the product in skin care during radiotherapy—was analyzed using instrumental–sensory analysis. The general research procedure is presented in Figure 1.

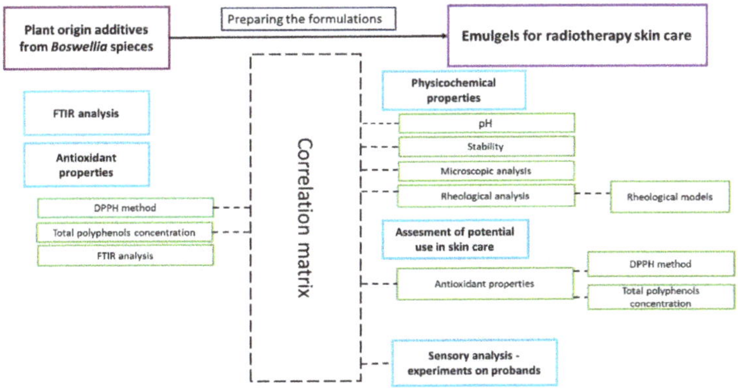

Figure 1. General research procedure.

2.1. Materials

In this research, cosmetic raw materials, such as refined sunflower oil (EOL Polska Sp. z o.o., Szamotuły, Poland), milk thistle oil (unrefined) (VitaFarm, Lipno, Poland), glyceryl monostearate (Zrób Sobie Krem, Prochowice, Poland), vegetable glycerin anhydrous (CHEMPUR, Piekary Śląskie, Poland), cetyl alcohol (Zrób Sobie Krem, Prochowice, Poland), D-panthenol (Zrób Sobie Krem, Prochowice, Poland), Euxyl PE 9010 (INCI: Phenoxyethanol (and) Ethylhexylglycerin) (Ashland Specialties Poland Sp. z o.o, Warszawa, Poland), essential oil from *Boswellia carterii* Birdw. (magiczneindie.pl, Warszawa, Poland), select-extract CO_2 (*Boswellia serrata* Roxb.) (Ecospa, Warszawa, Poland), powdered frankincense (*B. serrata*) (Nanga, Błękwit, Poland), and carbomer (Zrób Sobie Krem, Prochowice, Poland), were used. The chemical reagents used in the assessment of antioxidant properties were methanol anhydrous 99.8%, ethanol 96%, 2,2-diphenyl-1-picrylhydrazyl (DPPH), sodium bicarbonate, and Folin–Ciocalteu reagent (CHEMPUR, Piekary Śląskie, Poland).

2.2. *Characteristics of Extracts and Essential Oil from Boswellia Species*

Three types of plant raw materials obtained from the *Boswellia* family were used in this study. The ethanol extract from *B. serrata* was obtained by extracting powdered frankincense with 96% ethanol in a Soxhlet apparatus for 20 h. The obtained extract was a light yellow liquid with an intense odor. The remaining two raw materials, i.e., selective CO_2 extract and essential oil, were commercial products.

Each of the plant raw materials was subjected to Fourier transform infrared spectroscopy (FTIR) analysis and assessment of its antioxidant properties and polyphenol content.

2.2.1. FTIR Spectrum

FTIR spectra in a wavenumber range from 4000 cm^{-1} do 400 cm^{-1}, at resolution 1 cm^{-1}, were recorded at 22 °C on an FTS−165 spectrophotometer (FTIR Biorad, Krefeld, Germany). For each spectrum, 125 scans were registered. Pure plant raw materials were used in this study.

2.2.2. Antioxidant Activity by 1,1-Diphenyl-2-Picrylhydrazyl (DPPH) Assay

The free radical scavenging activity of plant-origin additives from *Boswellia* was determined using the DPPH method. Three cm^3 of methanolic DPPH solution (2.8 mg/50 cm^3) was added to 0.4 cm^3 of analyzed extract/essential oil or water (in the case of a blank sample) and were stored in dark place. After 30 min, the absorbance at a wavelength of λ = 517 nm was measured on the NANOCOLOR UV–Vis spectrophotometer (Macherey-Nagel, GmbH & Co, KG, Düren, Germany). Absorbance measurements were repeated three times. The inhibition percentage was calculated using Formula (1):

$$\%Inhibition = \frac{A_0 - A_{sample}}{A_0} \cdot 100\%, \quad (1)$$

where:

A_0—absorbance of blank sample;
A_{sample}—average absorbance of samples.

2.2.3. Total Phenolic Content Assay

The Folin–Ciocalteu reagent (diluted 10 times) was used to determine the total phenolic content in analyzed extract or essential oil from *Boswellia* spp. Five cm^3 of Folin–Ciocalteu reagent was added to 1 cm^3 of extract or essential oil (or 1 cm^3 of water in case of blank sample). After 4 min, 4 cm^3 t sodium bicarbonate solution (7.5% m/m) was added. The mixture obtained in this way were stored at 22 °C in dark place for another 2 h. The absorbance values at λ = 765 nm were measured using a NANOCOLOR UV-Vis (Macherey-Nagel, GmbH & Co, KG, Düren, Germany) spectrophotometer. Each measurement was repeated three times, and the absorbance values obtained were averaged. Based on the calibration curve for gallic acid, the obtained results were presented as the concentration of gallic acid (mg/dm^3).

2.3. *Composition and Obtaining of Topical Formulations*

A detailed recipe of the obtained formulations is presented in Table 1. Plant-origin additives from the *Boswellia* species were introduced into both the hydrogel and emulsion recipes. For comparative purposes, placebo systems were also created, i.e., formulations without active substances originating from the *Boswellia* spp.

Table 1. Recipe of topical formulations with plant-origin additives from Boswellia species. E—emulsion; Em—emulgel; P—placebo; CO_2—addition of CO_2 extract; O—addition of essential oil; S—addition of Soxhlet extract; h—plant raw material was introduced to the hydrogel.

Ingredient	E_P	Em_P	E_SCO_2	Em_SCO_2	E_SO	Em_SO	Em_hSCO_2	Em_hSO
	% Mass.							
Distilled water	59.60	29.80	56.10	28.05	55.10	27.55	28.05	27.55
Hydrogel	-	50.00	-	50.00	-	50.00	50.00	50.00
Sunflower oil	10.00	5.00	10.00	5.00	10.00	5.00	5.00	5.00
Milk thistle oil	10.00	5.00	10.00	5.00	10.00	5.00	5.00	5.00
Glyceryl monostearate	8.00	4.00	8.00	4.00	8.00	4.00	4.00	4.00
Glycerin	7.00	3.50	7.00	3.50	7.00	3.50	3.50	3.50
Cetyl alcohol	3.00	1.50	3.00	1.50	3.00	1.50	1.50	1.50
Soxhlet extract from *B. serrata*	-	-	2.50	1.25	2.50	1.25	1.25	1.25
Select CO_2 extract from *B. serrata*	-	-	1.00	0.50	-	-	0.50	-
Essential oil from *B. carterii*	-	-	-	-	2.00	1.00	-	1.00
D-panthenol	2.00	1.0	2.00	1.00	2.00	1.00	1.00	1.00
Mixture of phenoxyethanol and ethylhexylglycerin	0.40	0.20	0.40	0.20	0.40	0.20	0.20	0.20

2.3.1. Emulsion

The ingredients of the oil phase (sunflower oil, milk thistle oil, and cetyl alcohol) were melted in a water bath. The aqueous phase (distilled water, vegetable glycerin, and preservative) was heated to 60 °C. When both phases had the same temperature (approx. 60 °C), the oil phase was poured into the water phase in small portions, stirring intensively with an IKA RW 20 digital 450 rpm mechanical mixer (IKA Poland Sp. z o.o., Warsaw, Poland). After combining the phases, mixing was continued for 10 min and the cooling stage was started. After cooling the mixture to 40 °C, D-panthenol was added. The whole mixture was mixed until a homogeneous consistency and room temperature were obtained. In the case of emulsions containing *Boswellia* spp. extracts, the Soxhlet extract was added to the water phase of the emulsion and, when hot, combined with the oil phase, whereas the CO_2 extract—or, respectively, the essential oil—was added to the obtained emulsion at a temperature below of 40 °C.

2.3.2. Emulgel

A 1% carbomer solution was obtained using a Sunlab SU1150 450 rpm magnetic stirrer (Bionovo, Legnica, Poland) by dispersing carbomer in distilled water. Subsequently, 3–4 drops of triethanolamine were added to adjust the pH, which led to a gel consistency. *B. serrata* extracts and/or essential oil were added to the hydrogel at room temperature. The obtained emulsion and hydrogel were mixed with a mechanical mixer at 450 rpm at 25 °C in a weight ratio of 1:1 until a homogeneous mixture was obtained.

2.4. Physicochemical Analysis of Formulations

2.4.1. Visual Assessment and pH

In order to check the quality of the prepared formulations and possible changes in their visual appearance after adding plant-origin raw materials, they were visually assessed by a qualified person during quality control. The visual assessment concerned odor, color, consistency, and homogeneity. A total of 5 g of the analyzed sample was placed in the weighing dish and hermetically closed. During the evaluation, the container was opened, and the odor, color, and homogeneity of the cosmetic sample were evaluated. To evaluate consistency of sample, a finger was dipped in the formulation at an angle of 45–60° and pulled out quickly. During the test, it is necessary to pay attention to sample resistance when dipping the finger into the formulation and to the contact of the finger and the cosmetic when pulling it out.

The pH of the preparations was determined by immersing the measuring electrode of a pH meter (Ezodo 7200, Chemland, Stargard, Poland) directly in the preparation. The measurement was repeated three times. The obtained results constitute the arithmetic mean of the obtained results.

2.4.2. Stability

In order to confirm that the obtained formulations are stable, accelerated aging tests (variable temperature test) and long-term stability tests were conducted. The variable temperature test included three cycles of 24 h at each temperature (4/40 °C) for a single sample (Cultura® M 70708 Mini Incubator, Almedica AG, Galmiz, Switzerland). After this time, we assessed whether there were any unfavorable changes in color, odor, and consistency in the sample and whether phase separation occurred. Moreover, the prepared preparations were stored for 3 months at 4 °C and 25 °C in ambient humidity.

2.4.3. Microscopic Structure

The microscopic structures of the obtained emulsions and emulgels were determined under a Motic Advanced B1 optical microscope (Motic Asia, Hong Kong, China) at 40× magnification. To determine the mean particle diameter of the dispersed phase, Motic Images Advances 3.2 software was used.

2.4.4. Rheological Properties

For the obtained formulations, a flow curve and viscosity curve (shear rate range 1–1000 s^{-1}, 60 s, 60 measurement points) were determined using rotational rheometer RS Plus Brookfield (LaboPlus, Warszawa, Poland). The measurements system was cone–plate C25-2. The presented test results are the arithmetic means of the three measurements. The obtained rheological data results were approximated with the Herschel–Bulkley model:

$$\tau = \tau_0 + k \cdot \dot{\gamma}^n, \tag{2}$$

where:

τ_0—yield stress, Pa;
τ—shear stress, Pa;
k—consistency index, Pa·s^n;
$\dot{\gamma}$—shear rate, s^{-1};
n—flow index, -.

2.5. Antioxidant Properties and Polyphenols Content

To determine antioxidant properties and polyphenols content in the preparations, 1 g of each sample was dissolved in 5 cm^3 methanol then centrifuged for 15 min at 3500 rpm in a Hettich EBA 200 S centrifuge (Marazet, Poznan, Poland). The obtained filtrate was analyzed spectrometrically via the DPPH method and with Folin–Ciocalteu reagent. The same methodology as for the analysis of the plant extracts and essential oils was used.

2.6. Sensory Analysis

Sensory analysis for all obtained formulations was performed in a cosmetic laboratory. The oncocosmetic evaluation was carried out by persons trained in cosmetic product evaluations. The test involved 7 testers (participants): women with healthy skin, not undergoing radiotherapy treatment, aged 22–41. The participants were familiarized with the research procedures and the assessment of individual features before the assessment. The research was carried out using a scale of 1–5 points, where 1 is the lowest rating. The following features of the preparations were assessed: odor, consistency, cushion effect, spreadability, oiliness, absorption, and hydration. The detailed research procedures and definitions of the assessed features is presented in our previous publication [24].

All participants gave their informed consent for inclusion before they participated in this study. The scope of this study is in line with Regulation (EC) No 1223/2009 of the European Parliament and Council; Cosmetics Europe—The Personal Care Association Guidelines: "Product Test Guidelines for the Assessment of Human Skin Compatibility 1997"; Cosmetics Europe—The Personal Care Association: "Guidelines for the Evaluation of the Efficacy of Cosmetic Products 2008"; and the World Medical Association (WMA) Declaration of Helsinki's ethical principles for medical research involving human subjects.

2.7. Statistical Analysis

Whenever applicable, the obtained results were expressed as mean ± standard deviation. Analyses of variance (ANOVA) were conducted to check differences among pH, polyphenol concentrations, % of inhibition, and the sensory attributes of the obtained formulations. A value of $p < 0.05$ was considered statistically significant. In case of significant differences at 95%, Tukey's HSD post hoc comparison tests were then carried out. To show these differences, lowercase letters of the alphabet were used. Samples marked with different letters are statistically significantly different. Samples with the same letter do not differ significantly from each other. Samples with mixed letters (e.g., "ab") do not differ significantly from any of the samples marked with the letters "a" and "b".

The correlation matrix was used in order to establish and interpret the relationship between sensory and instrumental measurements.

3. Results

3.1. Physicochemical Properties of Obtained Formulations

The physicochemical properties of the obtained topical formulations are given in Table 2.

Table 2. Physicochemical properties of obtained topical formulations.

Sample	Odor	Color	Consistency	Stability	pH	Mean Droplet Size of Dispersed Phase µm
E_P	none	white	creamy	stable	6.69 ± 0.01 a	19.0 ± 1.5
Em_P	none	white	cream–gel	stable	5.88 ± 0.03 b	10.9 ± 0.5
E_SCO$_2$	intense herbal scent	white	creamy	stable	5.64 ± 0.01 bc	8.1 ± 0.6
Em_SCO$_2$	intense herbal scent	white	creamy	stable	6.12 ± 0.01 ab	15.1 ± 0.9
E_SO	intense herbal scent	white	creamy	stable	4.92 ± 0.06 cd	6.6 ± 0.7
Em_SO	intense herbal scent	white	creamy	stable	4.57 ± 0.03 cd	8.6 ± 0.3
Em_hSCO$_2$	intense herbal scent	white	creamy	stable	5.51 ± 0.01 cd	20.9 ± 0.8
Em_hSO	intense herbal scent	white	creamy	stable	4.77 ± 0.03 d	7.0 ± 0.6

E—emulsion; Em—emulgel; P—placebo; CO_2—addition of CO_2 extract; O—addition of essential oil; S—addition of Soxhlet extract; h—plant raw material was introduced to the hydrogel. Different letters in the same category indicate significant difference between samples at $p < 0.05$.

Under the analyzed storage conditions, the obtained samples were stable; no changes in their color, consistency, or odor were noted. There was also no phase separation. Samples E_P and Em_P, without active substances from the *Boswellia* spp., were white in color, had a uniform, smooth, creamy or cream–gel consistency, respectively, and were odorless.

After adding plant raw materials from the *Boswellia* spp. to the product recipe, there were no changes in their color or consistency. There was a change in the odor, which became intense, and a herbal scent was noticed. The pH of the obtained formulations was 4.57 ± 0.03–6.69 ± 0.01. Preparations E_SO, EmhSO, and Em_SO, with the addition of *B. carterii* essential oil, had a lower pH than Em_SCO$_2$ (with Soxhlet extract and CO$_2$ extract) and placebo formulations. Moreover, it has been shown that the method of the addition of the *Boswellia* raw material (either to the emulsion or to the hydrogel) did not affect the pH of the formulation.

Microscopic images of the preparations confirmed the structure of the oil in water macroemulgel (o/w macroemulgel) (Figure 2), where the continuous phase is a hydrogel matrix, and the dispersed phase is an oil-in-water emulsion (o/w emulsion). The structure of the emulgel was also preserved when the extract and/or oil were added to the hydrogel rather than to the emulsion. The obtained formulations were characterized by different droplet sizes of the dispersed phase: 6.6 ± 0.7–20.9 ± 0.8 μm.

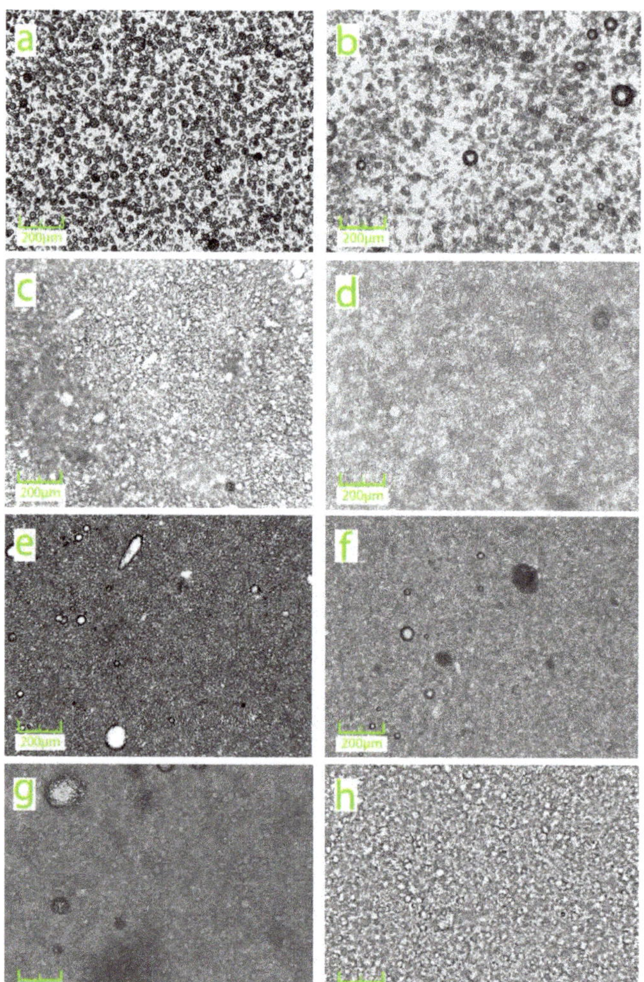

Figure 2. Microscopic images of the obtained formulations: (**a**) E_P; (**b**) Em_P; (**c**) E_SCO$_2$; (**d**) Em_SCO$_2$; (**e**) E_SO; (**f**) Em_SO; (**g**) Em_hSCO$_2$; (**h**) Em_hSO.

3.2. Rheological Properties

Figure 3 shows the flow curves of the prepared formulations. The flow curves were approximated by the Herschel–Bulkley model (Table 3).

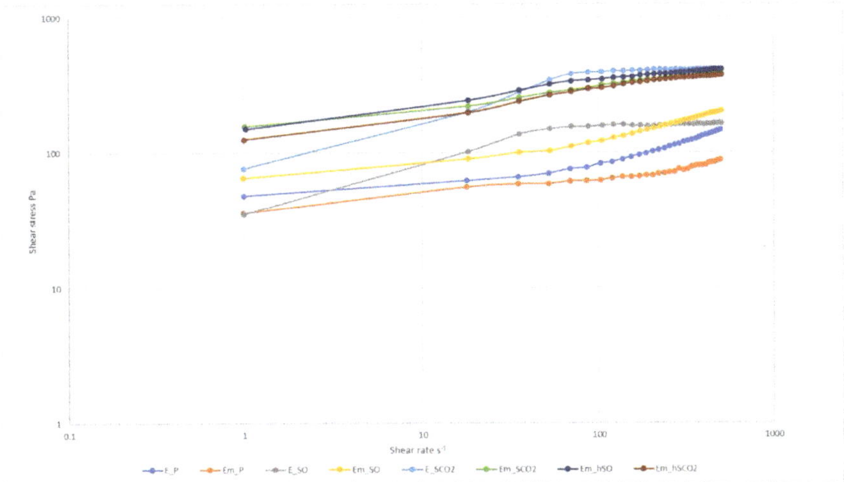

Figure 3. Flow curves of obtained samples. E—emulsion; Em—emulgel; P—placebo; CO_2—addition of CO_2 extract; O—addition of essential oil; S—addition of Soxhlet extract; h—plant raw material was introduced to the hydrogel.

Table 3. Parameters of the Herschel–Bulkley model and viscosity of samples.

Sample	τ_0, Pa	k, Pa·s^n	n, -	R^2	η, Pas at $\dot{\gamma}$ = 50 s^{-1}
E_P	49.27	1.372	0.686	0.999	1.35 ± 0.04
Em_P	36.03	4.629	0.378	0.976	1.20 ± 0.08
E_SO	16.96	66.53	0.136	0.925	3.21 ± 0.20
Em_SO	61.44	5.291	0.528	0.999	1.95 ± 0.12
E_SCO_2	24.07	154.3	0.161	0.929	6.50 ± 0.58
Em_SCO_2	122.6	57.19	0.247	0.990	5.62 ± 0.23
Em_hSO	116.8	80.47	0.214	0.982	6.05 ± 0.43
Em_$hSCO_2$	84.90	70.22	0.234	0.986	5.03 ± 0.30

E—emulsion; Em—emulgel; P—placebo; CO_2—addition of CO_2 extract; O—addition of essential oil; S—addition of Soxhlet extract; h—plant raw material was introduced to the hydrogel.

The obtained samples belonged to the group of non-Newtonian fluids shear thinning with a yield point, which, in the case of emulgels, results from the use of carbomer as a hydrogelator. Emulgels with the addition of *Boswellia* raw materials had a higher yield point than emulsions. Among them, the highest yield points were those of emulgels in which plant-origin additives were introduced into the hydrogel, not into the emulsion. The Em_P sample had the lowest viscosity, and E_SCO_2 had the highest. Moreover, despite the constant mass ratio of emulsion to hydrogel in the obtained emulgels, they differed in viscosity. The type of shear thinning flow property for the obtained samples can be described by the Herschel–Bulkley model (R^2 > 0.980).

3.3. Antioxidant Properties and Polyphenols Content of Formulations

Table 4 lists the types of the most important vibrations and the corresponding wavenumber for the tested plant materials of the *Boswellia* spp. The FTIR spectra of extracts from *Boswellia* spp. showed characteristic peaks at 3357 cm^{-1} and 3578 cm^{-1} (OH stretching); 2969 cm^{-1} and 2957 cm^{-1} (C-H stretching); 1712 cm^{-1} and 1642 cm^{-1} (C=O stretching); 1453 cm^{-1} and1381 cm^{-1} and at 879 cm^{-1} and 778 cm^{-1} (OH deformation), respectively, for Soxhlet and the select CO_2 extract. For essential oil from *B. carterii*, characteristic peaks were at 3446 cm^{-1} (OH stretching), 2957 cm^{-1} (C-H stretching), 1738 cm^{-1} (C=O stretching), 1447 cm^{-1}, and 778 cm^{-1} (OH deformation).

Because the intensity of the bands depended on the type of extraction method used, the *Boswellia* spp. raw materials used in this research were characterized by different concentrations of biologically active compounds, which translates in their different antioxidant activity and polyphenol contents (Figure 4).

The select CO_2 extract and essential oil (OE) had the highest % inhibition of the DPPH radical, amounting to 54.8%, and 49.9% respectively, while the essential oil had the highest polyphenol content of 57.3 mg/dm^3, expressed as gallic acid. α-pinene, α-thujene, β-pinene, and limonene may be responsible for such a high content of polyphenols, as well as the high % inhibition of the DPPH radical, while the high % inhibition in the case of the CO_2 extract may be due to the presence of *Boswellia serrata* resin extract, as well as compounds such as limonene and linalool, which have proven antioxidant properties. Although the essential oil itself had a higher polyphenol content, among the formulations, the highest concentration of polyphenols were had by sample E_SCO$_2$ (3.88 mg/dm^3) and sample Em_SO (3.64 mg/dm^3). This may be related to the fact that in recipe of emulsion, one of the components is milk thistle oil, which has polyphenols in its composition [25]. When comparing the % inhibition of the preparations, the E_SO containing essential oil had the highest value (18.5%). All preparations—except Em_hSO (its % inhibition of the DPPH radical was 7.29)—showed a higher % inhibition of the DPPH radical compared to the placebo preparations. In the case of polyphenol content, the statistical differences was shown for placebo formulation and samples: E_SCO$_2$, E_SO, Em_SO, and Em_hSO. This means that the addition of extracts and essential oil increased the antioxidant activity and polyphenol content in preparations containing plant-origin additives from the *Boswellia* species.

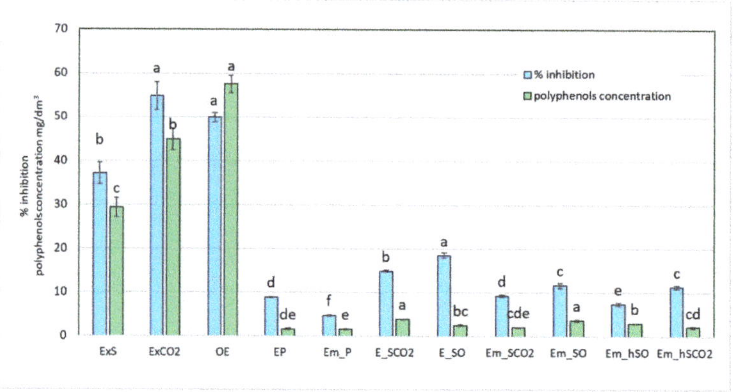

Figure 4. Results of antioxidant activity and polyphenols concentration in plant-origin additives from *Boswellia* species and topical formulations for skin care. ExS—Soxhlet extract; ExCO$_2$—CO_2 extract; OE—essential oil; E—emulsion; Em—emulgel; P—placebo; CO_2—addition of CO_2 extract; O—addition of essential oil; S—addition of Soxhlet extract; h—plant raw material was introduced to the hydrogel. Different letters in the same category (separately for raw materials and separately for preparations) indicate significant difference between samples at $p < 0.05$.

Table 4. Types of vibrations and corresponding wavenumber for analyzed plant-origin additives from *Boswellia* species. ν—stretching vibrations; δ—deformation vibrations.

Sample	Type of Vibrations and Corresponding Wavenumber cm^{-1}								
	ν-OH Alcohols, Phenols 3000–3500 cm^{-1} Wide Band at 3500–3550 cm^{-1} for Carboxylic Acids	ν C-H Aromatic 3030 cm^{-1}	ν C-H Aliphatic Compound 2850–3000 cm^{-1}	ν C=O Aldehydes, Ketones, Acids, Esters 1600–1870 cm^{-1}	ν C=C in Alkenes 1600–1680 cm^{-1}	ν C=C Aromatic 1500–1610 cm^{-1}	δ OH Phenols, Alcohols 1330–1420 cm^{-1}	ν C-O 1050–1430 cm^{-1}	δ OH Phenols Alcohols 650–770 cm^{-1} or ν C-Cl 600–800 cm^{-1}
Soxhlet extract	3357.19	-	2969.41	1712.12	-	-	1453.57 1377.26	1046.90	879.38
Select CO_2 extract	3578.17	3042.48	2957.32	1642.94	1642.94	1511.20	1381.69	1242.79	778.44
Essential oil from *B. carterii*	3446.06	3043.90	2957.74	1738.03	1643.38	1511.52	1447.08 1381.46	1242.57	778.30

3.4. Sensory Analysis

Results of the sensory analysis are presented in Figure 5.

The ANOVA statistical analysis did not show statistically significant differences for parameters such as odor, consistency, and hydration. The low scores in the odor category (the highest score was 3.29 ± 0.88 points) are related to the intense herbal smell of the used extracts and essential oil. Emulgels are known from their lighter consistency compared to emulsion, but in this case, the consistency of both types of formulations was rated good; the average score was 3.86–4.3 points. Also, the hydration for all preparations (even the placebo) was rated good; the averages score was 3.86–4.71 points. The average score for the cushion effect was in the range of 1.71–3.71 points. The statistical differences were noted for emulsions E_SCO$_2$ and E_P and emulgels Em_P, Em_hSCO$_2$, and Em_hSO. The increase in the yield point did not have a negative impact on the spreadability of the preparations. The lowest score in this category was had by the placebo emulsion (3.86 points), which was confirmed by the ANOVA statistical results. The most diverse assessments concerned the oiliness of the preparation (average score 1.0–3.86 points). Statistical differences were noted for placebo emulsion (E_P) and emulgels, namely, Em_P, Em_SO, Em_hSO, and Em_hSCO$_2$, which indicates that emulgels are less greasy than emulsions. The E_P formulation was rated the worst, which results from the fact that it is a placebo emulsion without the addition of any active substances. Statistical differences were also demonstrated for the placebo emulgel (Em_P) and the following samples: E_P, E_SO, Em_SO, E_SCO$_2$, Em_SCO$_2$, and Em_hSCO$_2$. The best rated in this category was Em_P, based on E_P. Statistically significant differences in absorption were observed between the placebo emulsion (E_P) and the emulgels in which the extract and/or essential oil were introduced into the hydrogel.

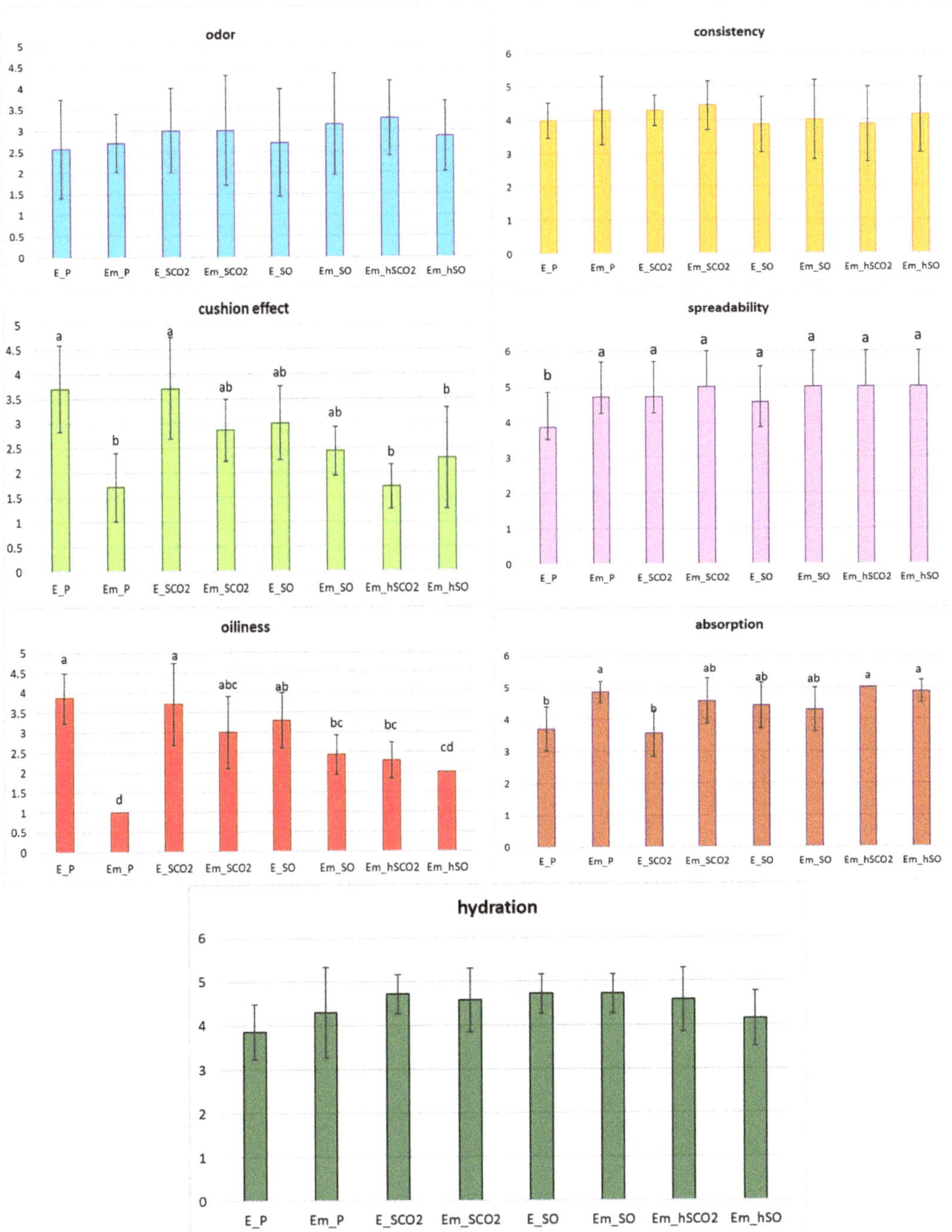

Figure 5. Sensory evaluation of obtained preparations. Different letters in the same category indicate significant differences between samples at $p < 0.05$. E—emulsion; Em—emulgel; P—placebo; CO_2—addition of CO_2 extract; O—addition of essential oil; S—addition of Soxhlet extract; h—plant raw material was introduced to the hydrogel.

4. Discussion

4.1. Composition and Characterization of Formulations

The clinical success in the antioxidants studied in radiodermatitis achieved with topical superoxide dismutase has stimulated interest in exploring naturally occurring and more-affordable antioxidant substances, especially those rich in polyphenols [10,26]. One of the plants that possesses therapeutic components is *Boswellia*. There are four main species of *Boswellia*, which include *Boswellia carterii*, *B. frerana*, *B. sacra*, and *B. serrata* [27].

This work is the first study indicating the possibility of using raw materials from *Boswellia* spp. as an antioxidant in topical preparations, i.e., emulgels and emulsions intended to care for the skin and limit unwanted skin reactions during radiotherapy.

Apart from being used in medicine, *Boswellia* is also an important cosmetic ingredient. There are 25 cosmetic ingredients in the CosIng database that can be obtained from the *Boswellia* species. These are resin extract, resin water, bark powder, gums, gum water, oils, and gum ferment extract.

Our research used select CO_2 extracts obtained via the extraction of supercritical carbon dioxide, which works as a solvent for naturally occurring plant ingredients. The CO_2 select extracts contain mainly volatile components typical for essential oils (with molecules ranging from C5 to C25). They have intensive, perceptible odors; heavier waxes, resins and color compounds are generally not present in this raw material, as reflected in the visual assessment and sensory analysis of samples. The CO_2 extract used, according to the manufacturer's data (INCI name: Boswellia Serrata Resin Extract, alpha-Pinene, Limonene, beta-Pinene, alpha-Terpinene, Terphinolene, Linalool, Limonene), contains approx. 70–90% essential oil, sesquiterpenes, diterpene alcohols (serratol, incensol), and boswellic acid. The allowable concentration of the extract for use in cosmetics ranges from 0.1 to 1% and from 0.4 to 0.75% in facial cosmetics.

Another ingredient was an essential oil obtained via the steam distillation of oleoresin *Boswellia carterii*. According to the manufacturer, the raw material came from India. It is a dark yellow liquid with an intense odor. It contains, among other ingredients, α-pinene, limonene, α-thujene, linalool, and cithrol. An extract rich in carbohydrates, tannins, glycosides, terpenes, flavonoids, and alkaloids can be obtained via Soxhlet extraction [28,29]. According to the literature data, they exhibit skin care effects, including anti-aging (reducing lines and wrinkles), anti-acne (anti-bacterial, anti-fungal properties), hair care (preventing hair loss), and nail care effects; they smooth, tonify, and protect skin, as well as providing fragrance. Due to their antioxidant properties, they prevent the formation of post-sun discoloration and reduce redness [30].

On the European Union market, there are no products dedicated to post-radiotherapy skin containing plant-origin additives from *Boswellia* species. However, all products available on the market containing this plant raw material are intended for very dry, damaged skin that requires regeneration, hydration, and soothing. Moreover, the market offers products in various physicochemical forms, such as bar soap, shower gel, serum, cream or sheet mask; however, there are no emulgel formulations.

Emulgels belong to the polymeric gels group: semi-solid formulations obtained by immobilizing the external phase of the emulsion (o/w or w/o type) using suitable gelators [31]. Therefore, they exhibit the properties of gels, which are characterized by a faster and better release of the active ingredient compared to an ointment or cream base. Due to their physicochemical form, they are characterized by higher stability than emulsions. In these systems, the risk of phase separation and the creaming phenomenon that occurs in emulsions is reduced. In addition, they are simple and economical to produce. The disadvantages of emulgels may include the poor absorption of macromolecules through the skin and the appearance of air bubbles after the preparation process [32].

4.1.1. Visual Assessment and pH

In the literature, the use of *B. serrata extracts* in emulgels as topical delivery systems (drug) was studied by Chandak et al. [33]. The *Boswellia-serrata*-loaded emulgel (BS-loaded emulgel) was an anti-rheumatoid agent. Carbomer was used as a hydrogelator. The obtained formulations were creamy, with a smooth homogeneous appearance. Also in this study, the produced emulgels were characterized by a creamy consistency and white color, which is a desirable feature in preparations for sore, damaged skin [34].

The odor of the obtained formulations was intense and herbal. At a later stage of modifying the formulation's recipe, the odor should be slightly eliminated as it could disturb people using this preparation after radiotherapy, for whom strong odors may enhance the effect of nausea.

According to Auerswald et al. [35], the pH of skin subjected to radiotherapy is close to the pH of non-irradiated skin, whereas in the area of the treatment wound, the pH increases significantly and is higher (above 7.5) than that of non-irradiated skin (pH in the range of 4.5–6.7). pH-gradients play a crucial role in recruiting keratinocytes inwards from the wound edges. Also, the radiation-induced damage of epidermis and adnexal structures may impair the maintenance of an acidic pH in the stratum corneum. The alkalization of the stratum corneum following radiotherapy may impair skin barrier function and may also impair wound healing and predispose radiation-damaged skin to bacterial (especially *Staphylococcus aureus*) and fungal (*Candida albicans*) infections, which grow better at neutral pH [35]. Thus, it is important that formulations used in skin care after radiotherapy have an acidic pH [36], which is consistent with our results. The acidic pH of the prepared formulations containing plant-origin additives from *Boswellia* species is also consistent with the results of research by Chandak et al. [33]; the obtained emulgels had a pH of 6.3–7.09. In this case, *Boswellia serrata* extract was added to the aqueous phase at a concentration of 1%.

4.1.2. Microscopic Structure

The mean droplet size of the dispersed phase depended on the physicochemical form of the obtained formulations as well as on the addition and concentration of plant-origin additives from *Boswellia* species. The placebo emulgel (Em_P) has lower values of the mean droplet size of the dispersed phase compared to the placebo emulsion (E_P). The hydrogelator used in the study—polyacrylic polymer—effectively immobilizes the emulsion within the network structure, reduces the motion of the droplets, and prevents the droplets from colliding [37].

The situation is different for samples with the addition of raw materials from the *Boswellia* species. For emulgels in which the Soxhlet extract and essential oil or the CO_2 extract was added to the emulsion, an increase in the droplet size was observed compared to the emulsions from which they were prepared. For emulgels in which active substances from *Boswellia* were added to the hydrogel, the size of the dispersed phase droplets depended on the type of active substances used. The lowest values of the mean droplet size of the dispersed phase were recorded for formulations containing essential oil, both for emulsion systems and emulgels (regardless of whether it was added to a gelified emulsion or to a hydrogel). The changes in droplet sizes in systems belonging to polymeric gels when plant extracts have been added to the systems are consistent with our previous observations [38] and the literature data [39]. Additionally, according to Arshad et al. [40], the globule size analysis in this study indicated that the emulgel had a small globule size in the range desirable for topical products.

4.1.3. Rheological Properties

The shear-thinning behavior of the obtained samples affects the product's resistance to incompatibilities such as creaming, flocculation, or coalescence [41]. Additionally, shear thinning behavior is necessarily important for developing a viscous film over the skin as well as increasing the absorption of active molecules via the topical route [40]. The type of shear thinning flow property of the obtained samples can by described by the Herschel–Bulkley model. This is consistent with the research results published by Dong et al. [37] and Ergin et al. [42].

The consistency index (k), shown in Table 3, reveals higher values when plant-origin additives are added to the systems. A decrease in the value of the k parameter is observed when the emulsion is immobilized inside the hydrogel. This parameter is also directly related to the viscosity of samples; when the k parameter increases, the viscosity also increases. The values of flow index $n < 1$ confirm that these are shear-thinning fluids. The least-plastic system (n is the highest) is the placebo emulsion. The value of the flow index is influenced by the presence and type of plant raw material used, as in the research results obtained by Di Mambro [43]. The effect we observed of the addition of plant extract on the decrease/increase in the viscosity of the formulation is also consistent with the results published by Di Mambro [43]. Moreover, the observed changes in viscosity are influenced by the viscosity of the extract used and/or the presence of proteins in the extract [44]. Proteins spontaneously adsorb to oil–water interfaces, essentially due to the hydrophobic properties of these interfaces. Adsorption then results in the decrease in the interfacial tension and a formation of a large interface area in emulsion [43]. The high yield point in the Em_SCO$_2$, Em_hSO, Em_hSCO$_2$, Em_hSCO$_2$, and E_SCO$_2$ formulations did not have a negative impact on the spreadability assessed in the sensory analysis. This parameter was evaluated positively by the probands for all preparations. The average rating was in the range of 3.57–5.00.

4.2. Antioxidant Properties and Polyphenols Content

The chemical composition of *Boswellia* oleoresin depends on its botanical origin, while the type and concentration of compounds present in the extract depends on the extraction method used and the type of solvent [29]. According to the literature data, the oleoresin consists of carbohydrates, terpenoids, phytosterols, phenolic compounds, flavonoids, and tannins, including α- and β-boswellic acids belonging to pentacyclic triterpenes, which may constitute 30–60%, and essential oils amounting to 5–10% (with the main component of essential oils being α-pinene) [45,46]. The data presented in Table 4 prove that the tested extracts and essential oils of the *Boswellia* species contain compounds from the following groups: alcohols, phenols, carboxylic acids, including boswellic acid, esters, alkenes, alkanes, aliphatic amines, halides, cycloalkanes, which is consistent with literature data [47–50].

The pure extract samples and essential oil tested in the present study for their H-donor ability, measured by the stable free radical DPPH$^•$ assay, showed a higher antioxidant activity when compared to formulation samples. The content of polyphenols for plant raw materials is also higher than for the tested preparations. According to [51], the reason for the much lower antioxidant activity and polyphenol content for the preparations compared to pure plant raw materials may be adding of the Soxhlet extract to the formulation at a high temperature, which could have caused a decrease in its activity as a result of the thermal degradation of some antioxidant compounds, such as phenolic compounds, that are present in the extract. A negative effect of heat on the stability of some polyphenolic antioxidants was also previously described in [52,53]. However, according to Di Mambro [43], these differences may result from the assay method used. In the author's opinion, the chemiluminescence assay shows a similar activity profile for extracts and samples, which cannot be observed in the DPPH analysis. These results may be due to the presence of the formulation components in the reaction mixture. The formulation components may interfere with the antioxidant measurements.

4.3. Correlation Matrix of Instrumental Analysis Results and Sensory Analysis

The use of sensory analysis to determine not only the maximum concentration but also the impact of the addition of plant extracts on the sensory characteristics of a product—while maintaining its antioxidant properties without deteriorating these characteristics, which as important for consumers—is being studied in the food industry [54,55]. This is due to the fact that plant extracts, especially those rich in phenolic compounds, may have taste, color, and odor, and all such aspects should be taken into consideration prior to any use in the formulation of foods.

In the case of cosmetic and pharmaceutical preparations, sensory characteristics such as smell, consistency, spreadability, and greasiness after application are important to consumers, especially for people with skin diseases or injuries [32,35]. When it comes to assessing the antioxidant capacity of topical preparations containing plant extracts in their formulations, the authors approach the issue in two ways. The first method involves evaluating the extracts and the prepared formulations separately.

The extracts are analyzed for their antioxidant capacity (usually via the DPPH or FRAP method), and their chemical composition is determined. The release of phenolic compounds from the gel matrix is also studied [56]. For the formulations, however, in vivo tests on probands are carried out, which include instrumental tests on the skin concerning skin melanin level, skin erythema, skin moisture content, skin sebum content, and skin elasticity [57].

The second method involves the assessment of the antioxidant potential of extracts and formulations using the same in vitro methods, although this approach is less popular in the literature. An example of such research is the work of Di Mambro et al. [43], where emulsions containing plant extracts for antioxidant activity tests were diluted 1:2.5 in a suitable buffer for each assay. All diluted formulations were mixed for 20 min prior to the measurements of their antioxidant activity.

In this research, the second approach was used, i.e., the assessment of antioxidant properties for both plant raw materials and formulations. For the emulsions and emulgels, the in vitro method of obtaining filtrates proposed in studies of bigels [38] containing *Centella asiatica* was used.

Additionally, an attempt to correlate the impact of plant-origin additives on the product functional properties determined via instrumental methods with the sensory characteristics determined in the sensory analysis was made.

In addition to the basic physicochemical characteristics, the analyzed functional properties included those regarding the effectiveness of the preparation as a potential product in post-radiotherapy skin care, i.e., % inhibition and the content of polyphenols in the formulation composition. The obtained results are presented in the form of a correlation matrix in Table 5.

The presence of an essential oil and/or a select CO_2 extract rich in volatile ingredients (such terpenoids, as well as aliphatic octyl acetate [44]) in the formulation influences the product's odor, which is correlated with hydration (correlation coefficient 0.625 $p \geq 0.1$) and spreadability (correlation coefficient 0.760 $p \geq 0.05$). The literature data [30,44] indicate the presence of moisturizing ingredients such as polysaccharides, proteins, and boswellic acids, which help maintain the proper level of hydration. *Boswellia* species also have a proven anti-inflammatory effect, which affects the condition of dry, irritated skin, and skin with dermatoses [58]. In addition, it has been reported that *Boswellia* can be an effective topical agent with which to soften facial lines and relax the skin [59]. However, in the case of the correlation between odor and spreadability, it can be presumed that the probands spread the formulation with an intense odor faster more effectively and also more willingly. The odor dissipates faster during vigorous movements. Moreover, according to Mun et al., the presence of thickening substances in the system, especially carbomer in emulgels and cetyl alcohol in emulsions, can affect the perception of odor by the probands [60]. The cetyl alcohol present in the formulation of products also enhances the odor of the product and perfumes the skin.

Table 5. Correlation matrix of the results of physicochemical, rheological, sensory, and functional properties of formulations. * significance at $p < 0.001$; ** significance at $p < 0.01$; *** significance at $p < 0.05$; **** significance at $p < 0.1$.

Correlations	Odor	Consistency	Cushion Effect	Spreadability	Oiliness	Absorption	Hydration	% Inhibition	Polyphenol Content	pH	Mean Droplet Size	Viscosity	Yield Point	k	n
Odor	1.00														
Consistency	−0.07	1.00													
Cushion effect	−0.39	0.11	1.00												
Spreadability	0.760 ***	0.22	−0.62	1.00											
Oiliness	−0.15	−0.14	0.920 **	−0.53	1.00										
Absorption	0.28	−0.09	−0.919 **	0.62	−0.808 ***	1.00									
Hydration	0.625 ****	−0.03	−0.08	0.60	0.07	0.04	1.00								
% inhibition	0.15	−0.45	0.45	−0.04	0.624 ****	−0.38	0.662 ****	1.00							
Polyphenol content	0.44	0.01	0.26	0.39	0.23	−0.37	0.58	0.50	1.00						
pH	−0.38	0.36	0.36	−0.62	0.30	−0.33	−0.51	−0.35	−0.644 ****	1.00					
Mean droplet size	0.18	−0.16	−0.09	−0.27	0.12	0.08	−0.34	−0.30	−0.628 ****	0.677 ****	1.00				
Viscosity	0.43	0.31	0.11	0.48	0.20	0.07	0.29	0.21	0.36	−0.12	−0.09	1.00			
Yield point	0.36	0.24	−0.36	0.49	−0.27	0.54	−0.21	−0.50	−0.20	−0.04	0.30	0.39	1.00		
k	0.45	0.39	0.22	0.35	0.22	−0.21	0.27	0.16	0.53	−0.03	−0.11	0.874 **	0.10	1.00	
n	−0.30	−0.14	0.15	−0.57	0.07	−0.33	−0.57	−0.43	−0.29	0.38	0.39	−0.764 ***	−0.05	−0.57	1.00

The strongest correlation (correlation coefficient closest to 1 $p \geq 0.01$) was obtained for the cushion effect and oiliness (0.920) and for the cushion effect and absorption (correlation coefficient −0.919). The cushion effect is the amount of product felt between the fingers (forefinger and thumb) when rubbing them together. The more product you feel between your fingers, the stronger the cushion effect. The strongest cushion effect was reported by the probands for emulsions and was smallest for emulgels, which is related to their physicochemical form. This effect is typical for emulsions and less noticeable for emulgels due to their excellent spreading properties combined with fast and easy absorption [61], hence the negative correlation observed; the smaller the cushion effect (lower numerical value), the higher the absorption (higher score from the probands), as less formulation between the fingers will result in faster absorption. The smaller the amount of preparation between the fingers, as in the case of emulgels, the less oiliness felt, i.e., the degree of oily deposit left on the skin immediately after application.

Oiliness is correlated with absorption (correlation coefficient −0.808 $p \geq 0.05$). Samples of preparations that are too viscous and oily in feel may have poorer absorption for the probands. Hence, the lowest oiliness values were recorded for emulgels. This is desirable because the skin affected by radiodermatitis is sore, and patients with dermatoses and skin injuries have a negative opinion of oily, sticky preparations such as ointments and creams that they have to use [32,34].

The obtained tests results also showed that oiliness is correlated with the % inhibition of the sample. The % inhibition of the sample value is related to the presence of active substances in its composition. They are capable of neutralizing free radicals, the main sources of which are essential oil and select CO_2 extract, with an inhibition capacity of 49.9% and 54.8%, respectively. These raw materials themselves are not greasy because they are volatile; the essential oil evaporates without leaving a greasy stain. However, in this research, they have been introduced into the recipe of emulsions and emulgels. Low oiliness was recorded for emulgels where the essential oil and select CO_2 extract were added to the hydrogel. Moreover, according to Tasneem et al. [57], in the case of carbomer-based emulgels containing plant extracts with antioxidants and lipid lowering activity, such as *Boswellia* [62], it is possible to decrease the production of sebum. The obtained research results are surprising, and it should be emphasized that correlations of sensory-assessed features with the results of physicochemical tests of preparations for topical use are not a common topic of scientific studies.

Hydration is also correlated with the % inhibition of the sample, as confirmed by the moisturizing effect of plant-origin additives from *Boswellia* species, mentioned earlier and proven in the literature. However, the polyphenol content was correlated with the pH of the formulation (correlation coefficient: −0.644, $p \geq 0.1$). This is a further confirmation of the research results by Spiegel et al. [63], i.e., that an increase in pH—and hence the deprotonation of hydroxyl groups—increases the antiradical activity of polyphenols.

The correlation coefficient of the polyphenol concentration with the mean droplet size of the dispersed phase was −0.628 ($p \geq 0.1$). As was shown in research on polymeric gels [38,39], the addition of essential oils/plant extracts affects the mean droplet size of the dispersed phase.

Also, according to the literature data [64,65], pH can be a factor influencing the mean droplet size of the dispersed phase. The correlation coefficient of sample pH with the mean droplet size of the dispersed phase in the analyzed systems was 0.677 ($p \geq 0.1$).

The viscosity of the sample is correlated with the consistency coefficient (0.874, $p \geq 0.01$) and the flow index (−0.764. $p \geq 0.01$). This is due to the nature of the flow of the obtained formulations, which is discussed in detail in the discussion section on the rheological properties of formulations. Formulations with a pseudoplastic flow produce a coherent film covering the skin surface, and this is important for better antioxidant protection of the skin surface.

The produced preparations had the desired characteristics of formulations intended for oncological patients undergoing radiotherapy as they meet the requirements of cosmetics intended for sensitive skin (i.e., they have the appropriate pH [66]); injured skin and skin painful to the touch (the physicochemical form of emulgel has an appropriate soft consistency, leaves a slightly greasy feeling on the skin, and has good lubricity [32,34]); and skin treated with ionizing radiation beams (the substances present in the composition, such as milk thistle oil and a mixture of extracts and essential oil from *Boswellia* species with antioxidant properties, play a key role in protecting the skin against the harmful effects of free radicals and thus accelerate the regeneration of the injured epidermis [2,10]).

5. Conclusions

The possibility of using plant extracts in the care of oncological patients, particularly in terms of preventing dermatoses accompanying radiotherapy, is a relatively new and current issue being addressed by the scientific community and by research and development departments of cosmetic and pharmaceutical companies. Plant-origin additives from *Boswellia* species exhibiting anti-inflammatory, antioxidant, antibacterial, and regenerative properties have great potential to become key ingredients in the formulation of oncocosmetics, especially in the form of emulgels absent from the European market.

The most promising preparation is the emulgel containing the Soxhlet extract and essential oil (Em_SO) due to its high antioxidant properties and polyphenol concentration compared to other preparations. Also in favor of the preparation is its physicochemical form, which may represent an advantage in terms of transporting the active substance deep into the skin, and its sensory characteristics, which were also positively rated by the probands.

Low ratings from the probands were awarded to the placebo preparations, particularly Em_P. The demonstrated correlations of odor with hydration and spreadability, oiliness and hydration with the % inhibition of the sample, and the content of polyphenols with the pH and the size of the dispersed phase droplets—all these factors proved the positive effect of the addition of plant-origin additives from *Boswellia* species to the emulgel formulation on functional and sensory properties.

The results of our research will enable the development and production of new, innovative topical dosage forms which can be used in radiodermatitis prevention and treatment.

Author Contributions: Conceptualization, A.K.-P.; methodology, A.K.-P.; formal analysis, A.K.-P.; W.G.; investigation, A.K.-P., W.G. and J.K.; resources, A.K.-P. and W.G.; writing—original draft preparation, A.K.-P., W.G., J.K. and A.P.; writing—review and editing, A.K.-P., W.G., J.K. and A.P.; visualization, A.K.-P. and W.G.; supervision, A.K.-P.; project administration, A.K.-P.; funding acquisition, J.K. All authors have read and agreed to the published version of the manuscript.

Funding: This research received no external funding.

Institutional Review Board Statement: The study was conducted in accordance with the Declaration of Helsinki. Ethical review and approval were waived for this study. According to the EU cosmetic Regulation no. 1223/2009, the cosmetic product must not cause damage to human health when applied under normal or reasonably foreseeable conditions of use and must be assessed for its safety of use before human subjects are exposed to it, and as such, further ethical approval is not required.

Informed Consent Statement: Informed consent was obtained from all subjects involved in the study.

Data Availability Statement: The authors confirm that the data supporting the findings of this study are available within the article.

Acknowledgments: The authors would like to thank Monika Korpak for providing language help.

Conflicts of Interest: The authors declare no conflicts of interest.

References

1. Zeman, E.M.; Schreiber, E.C.; Tepper, J.E. 27–Basics of Radiation Therapy. In *Abeloff's Clinical Oncology*, 6th ed.; Niederhuber, J.E., Armitage, J.O., Kastan, M.B., Doroshow, J.H., Tepper, J.E., Eds.; Elsevier: Amsterdam, The Netherlands, 2020; pp. 431–460.e3. [CrossRef]

2. Kulawik-Pióro, A.; Goździcka, W.J. Plant and Herbal Extracts as Ingredients of Topical Agents in the Prevention and Treatment Radiodermatitis: A Systematic Literature Review. *Cosmetics* **2022**, *9*, 63. [CrossRef]
3. Abdel-Wahab, M.; Gondhowiardjo, S.S.; Rosa, A.A.; Lievens, Y.; El-Haj, N.; Polo Rubio, J.A.; Prajogi, G.B.; Helgadottir, H.; Zubizarreta, E.; Meghzifene, A.; et al. Global Radiotherapy: Current Status and Future Directions-White Paper. *JCO Glob. Oncol.* **2021**, *7*, 827–842. [CrossRef] [PubMed]
4. Rucińska, M. Radiotherapy in the combined treatment. *Nowotw. J. Oncol.* **2022**, *7*, 366–372. [CrossRef]
5. Bensadoun, R.J.; Humbert, P.; Krutman, J.; Luger, T.; Triller, R.; Rougier, A.; Seite, S.; Dreno, B. Daily baseline skin care in the prevention, treatment, and supportive care of skin toxicity in oncology patients: Recommendations from a multinational expert panel. *Cancer Manag. Res.* **2013**, *5*, 401–408. [CrossRef]
6. Radiation Therapy Side Effects. Available online: https://www.cancer.org/treatment/treatments-and-side-effects/treatment-types/radiation/effects-on-different-parts-of-body.html (accessed on 29 April 2024).
7. Zasadziński, K.; Spałek, M.J.; Rutkowski, P. Modern Dressings in Prevention and Therapy of Acute and Chronic Radiation Dermatitis—A Literature Review. *Pharmaceutics* **2022**, *14*, 1204. [CrossRef]
8. Kiprian, D.; Szykut-Badaczewska, A.; Gradzińska, A.; Czuwara, J.; Rudnicka, L. How to manage radiation-induced dermatitis? *Nowotw. J. Oncol.* **2022**, *72*, 86–95. [CrossRef]
9. Michalewska, J. Odczyny popromienne w radioterapii oraz popromienne zapalenie skóry. *Lett. Oncol. Sci.* **2017**, *14*, 104–109. [CrossRef]
10. Amber, K.T.; Shiman, M.I.; Badiavas, E.V. The use of antioxidants in radiotherapy-induced skin toxicity. *Integr. Cancer Ther.* **2014**, *13*, 38–45. [CrossRef]
11. Kim, J.H.; Kolozsvary, A.J.; Jenrow, K.A.; Brown, S.L. Mechanism of radiation-induced skin injury and implications for future clinical trials. *Int. J. Radiat. Biol.* **2013**, *89*, 311–318. [CrossRef]
12. Fowble, B.; Yom, S.S.; Yuen, F.; Arron, S. *Skin Care in Radiation Oncology. A Practical Guide*; Springer: Cham, Switzerland, 2016; pp. 31–45.
13. Draelos, Z.D. The science behind skin care: Moisturizers. *J. Cosmet. Dermatol.* **2018**, *17*, 138–144. [CrossRef]
14. Purnamawati, S.; Indrastuti, N.; Danarti, R.; Saefudin, T. The Role of Moisturizers in Addressing Various Kinds of Dermatitis: A Review. *Clin. Med. Res.* **2017**, *15*, 75–87. [CrossRef] [PubMed]
15. Lodén, M. Role of topical emollients and moisturizers in the treatment of dry skin barrier disorders. *Am. J. Clin. Dermatol.* **2003**, *4*, 771–788. [CrossRef] [PubMed]
16. McQuestion, M. Evidence-based skin care management in radiation therapy. *Semin. Oncol. Nurs.* **2006**, *23*, 163–173. [CrossRef] [PubMed]
17. Kodiyan, J.; Amber, K.T. Topical antixodidants in radiodermatitis: A clinical review. *J. Palliat. Nurs.* **2015**, *21*, 446–452. [CrossRef] [PubMed]
18. Wszołek, K.; Piotrowska, A. Analysis of the composition of selected cosmetics for oncological patients. *Kosmetol. Estet.* **2019**, *8*, 575–580.
19. Msomi, N.Z.; Simelane, M.B.C. 11. Herbal Medicine. In *Herbal Medicine*; IntechOpen: London, UK, 2019. [CrossRef]
20. Maver, T.; Maver, U.; Kleinschek, K.S.; Smrke, D.M.; Kreft, S. A review of herbal medicines in wound healing. *Int. J. Dermatol.* **2015**, *54*, 740–751. [CrossRef]
21. Skalska-Kamińska, A.; Woźniak, A.; Paduch, R.; Kocjan, R.; Rejdak, R. Herbal preparation extract for skin after radiotherapy treatment. Part One—Preclinical tests. *Acta Pol. Pharm. Drug Res.* **2014**, *71*, 781–788.
22. Griñan-Lison, C.; Blaya-Cánovas, J.L.; López-Tejada, A.; Ávalos-Moreno, M.; Navarro-Ocón, A.; Cara, F.E.; González-González, A.; Lorente, J.A.; Marchal, J.A.; Granados-Principal, S. Antioxidants for the Treatment of Breast Cancer: Are We There Yet? *Antioxidants* **2021**, *10*, 205. [CrossRef]
23. Thornfeldt, C.R. Chapter 12: Rośliny jako kosmeceutyki. In *Kosmeceutyki*, 2nd ed.; Draelos, Z.D., Ed.; Elsevier Edra Urban & Partner: Wrocław, Poland, 2009; pp. 87–98.
24. Kulawik-Pióro, A.; Ptaszek, A.; Kruk, J. Effective tool for assessment of the quality of barrier creams–relationships between rheological, textural and sensory properties. *Regul. Toxicol. Pharmacol.* **2019**, *103*, 113–123. [CrossRef]
25. Kalinowska, M.; Płońska, A.; Trusiak, M.; Gołębiewska, E.; Gorlewska-Pietluszenko, A. Comparing the extraction methods, chemical composition, phenolic contents and antioxidant activity of edible oils from Cannabis sativa and Silybum marianu seeds. *Sci. Rep.* **2022**, *12*, 20609. [CrossRef]
26. Kwon, M.J.; Kim, B.; Lee, Y.S.; Kim, T.Y. Role of superoxide dismutase 3 in skin inflammation. *J. Dermatol. Sci.* **2012**, *67*, 81–87. [CrossRef] [PubMed]
27. Alotaibi, S.H. Biophysical properties and finger print in *Boswellia* sp. *Burseraceae. Saudi J. Biol. Sci.* **2019**, *26*, 1450–1457. [CrossRef] [PubMed]
28. Srujana, T.S.; Babu, K.R.; Rao, B.S.S. Phytochemical Investigation and Biological Activity of Leaves Extract of Plant Boswellia serrata. *Pharma. Innov.* **2012**, *1*, 22–46.
29. Mishra, S.; Bishnoi, R.S.; Maurya, R.; Jain, D. *Boswellia serrata* roxb.–A bioactive herbs with various pharmacological activities. *Asian J. Pharm. Clin. Res.* **2020**, *13*, 33–39. [CrossRef]
30. Alraddadi, B.G.; Shin, H.-J. Biochemical Properties and Cosmetic Uses of *Commiphora myrrha* and *Boswellia serrata*. *Cosmetics* **2022**, *9*, 119. [CrossRef]

31. Sharma, V.; Nayak, S.K.; Paul, S.R.; Choudhary, B.; Ray, S.S.; Pal, K. 9–Emulgels. In *Woodhead Publishing Series in Biomaterials, Polymeric Gels: Characterization, Properties and Biomedical Applications*; Woodhead Publishing: Sawston, UK, 2018; pp. 251–264. [CrossRef]
32. Kulawik-Pióro, A.; Miastkowska, M. Polymeric Gels and Their Application in the Treatment of Psoriasis Vulgaris: A Review. *Int. J. Mol. Sci.* **2021**, *22*, 5124. [CrossRef]
33. Chandak, S.M.; Trivedi, S.S.; Wadher, K.J.; Umekar, M. Desing, Development and Ex vivo Characterization of Boswellia serrata Loaded Emulgel. *Int. Pharm. Phytochem. Res.* **2020**, *12*, 78–84. [CrossRef]
34. Miastkowska, M.; Kulawik-Pióro, A.; Szczurek, M. Nanoemulsion Gel Formulation Optimization for Burn Wounds: Analysis of Rheological and Sensory Properties. *Processes* **2020**, *8*, 1416. [CrossRef]
35. Auerswald, S.; Schreml, S.; Meier, R.; Blancke Soares, A.; Niyazi, M.; Marschner, S.; Belka, C.; Canis, M.; Haubner, F. Wound monitoring of pH and oxygen in patients after radiation therapy. *Radiat. Oncol.* **2019**, *14*, 199. [CrossRef]
36. Hu, S.C.; Hou, M.F.; Luo, K.H.; Chuang, H.Y.; Wei, S.Y.; Chen, G.S.; Chiang, W.; Huang, C.J. Changes in biophysical properties of the skin following radiotherapy for breast cancer. *J. Dermatol.* **2014**, *41*, 1087–1094. [CrossRef]
37. Dong, L.; Liu, C.; Cun, D.; Fang, L. The effect of rheological behavior and microstructure of the emulgels on the release and permeation profiles of Terpinen-4-ol. *Eur. J. Pharm. Sci.* **2015**, *12*, 140–150. [CrossRef] [PubMed]
38. Kulawik-Pióro, A.; Osak, E.; Mendrycka, M.; Trześniewska-Ofiara, Z. Bigels as novel systems for the delivery active compounds from *Centella asiatica*. *Soft Mater.* **2023**, *21*, 316–338. [CrossRef]
39. Gallegos-Infante, J.A.; Galindo-Galindo, M.d.P.; Moreno-Jiménez, M.R.; Rocha-Guzmán, N.E.; González-Laredo, R.F. Effect of Aqueous Extracts of *Quercus resinosa* on the Mechanical Behavior of Bigels. *Sci. Pharm.* **2022**, *90*, 73. [CrossRef]
40. Arshad, W.; Khan, H.M.S.; Akhtar, N.; Mohammad, I.S. Polymeric emulgel carrying *Cinnamomum tamala* extract: Promising delivery system for potential topical applications. *Braz. J. Pharm. Sci.* **2020**, *56*, e18318. [CrossRef]
41. de Lafuente, Y.; Ochoa-Andrade, A.; Parente, M.E.; Palena, M.C.; Jimenez-Kairuz, A.F. Preparation and evaluation of caffeine bioadhesive emulgels for cosmetic applications based on formulation design using QbD tools. *Int. J. Cosmet. Sci.* **2020**, *42*, 548–556. [CrossRef]
42. Ergin, A.D.; Inal, O.; Barkat, A. In vitro and ex vivo assessments of surfactant-free topical curcumin emulgel. *J. Res. Pharm.* **2023**, *27*, 544–556. [CrossRef]
43. Di Mambro, V.M.; Fonseca, M.J. Assays of physical stability and antioxidant activity of a topical formulation added with different plant extracts. *J. Pharm. Biomed. Anal.* **2005**, *37*, 287–295. [CrossRef]
44. Al-Yasiry, A.; Kiczorowska, B. Frankincense—Therapeutic properties. *Adv. Hyg. Exp. Med.* **2016**, *70*, 380–391. [CrossRef]
45. Singh, H.P.; Yadav Kumar, I.; Chandra, D.; Jain, D.A. In Vitro Antioxidant and Free Radical Scavenging Activity of Different Extracts of *Boerhavia diffusa* and *Boswellia serrata*. *Int. J. Pharma Sci. Res.* **2012**, *3*, 503–511.
46. Siddiqui, M.Z. *Boswellia serrata*, a potential anti-inflammatory agent: An overview. *Indian J. Pharm. Sci.* **2011**, *73*, 255–261. [CrossRef]
47. Mehta, M.; Dureja, H.; Garg, M. Development and optimization of boswellic acid-loaded proniosomal gel. *Drug Deliv.* **2016**, *23*, 3072–3081. [CrossRef]
48. Kora, A.J.; Sashidhar, R.B.; Arunachalam, J. Aqueous extract of gum olibanum (*Boswellia serrata*): A reductant and stabilizer for the biosynthesis of antibacterial silver nanoparticles. *Process Biochem.* **2012**, *47*, 1516–1520. [CrossRef]
49. Geetha, V.; Chakravarthula, S.N. Chemical composition and anti-inflammatory activity of *Boswellia ovalifoliolata* essential oils from leaf and bark. *J. For. Res.* **2017**, *29*, 373–381. [CrossRef]
50. Fatimah, A.; Albasher, G.; Alharbi, R.I.; Asaggabi, N.S. Antifungal Potential of Aqueous Extract of *Boswellia carteri*. *J. Pure Appl. Microbiol.* **2019**, *13*, 2375–2381. [CrossRef]
51. Almeida, I.F.; Maleckova, J.; Saffi, R.; Monteiro, H.; Góios, F.; Amaral, M.H.; Costa, P.C.; Garrido, J.; Silva, P.; Pestana, N.; et al. Characterization of an antioxidant surfactant-free topical formulation containing *Castanea sativa* leaf extract. *Drug Dev. Ind. Pharm.* **2015**, *41*, 148–155. [CrossRef] [PubMed]
52. Maghsoudlou, Y.; Asghari Ghajari, M.; Tavasoli, S. Effects of heat treatment on the phenolic compounds and antioxidant capacity of quince fruit and its tisane's sensory properties. *J. Food Sci. Technol.* **2019**, *56*, 2365–2372. [CrossRef]
53. Ghafoor, K.; Mohamed, I.A.; Doğu, S.; Uslu, N.; Fadimu, G.J.; Al Juhaimi, F.; Babiker, E.; Özcan, M.M. The Effect of Heating Temperature on Total Phenolic Content, Antioxidant Activity, and Phenolic Compounds of Plum and Mahaleb Fruits. *Int. J. Food Eng.* **2019**, *15*, 20170302. [CrossRef]
54. Bernal-Mercado, A.T.; Ayala-Zavala, J.F.; Cruz-Valenzuela, M.R.; Gonzalez-Aguilar, G.A.; Nazzaro, F.; Fratianni, F.; de Miranda, M.R.A.; Silva-Espinoza, B.A. Using Sensory Evaluation to Determine the Highest Acceptable Concentration of Mango Seed Extract as Antibacterial and Antioxidant Agent in Fresh-Cut Mango. *Foods* **2018**, *7*, 120. [CrossRef]
55. McBride, N.; Hogan, S.A.; Kerry, J.P. Comparative addition of rosemary extract and additives on sensory and antioxidant properties of retail packaged beef. *Int. J. Food Sci. Technol.* **2007**, *42*, 1201–1207. [CrossRef]
56. Butkeviciute, A.; Ramanuskiene, K.; Janulis, V. Formulation of gels and Emulgels with Malus domestica Borkh: Apple extracts and Their Biopharmaceutical Evaluation In Vitro. *Antioxidants* **2022**, *11*, 372. [CrossRef]
57. Tasneem, R.; Khan, H.; Zaka, H.S.; Khan, P. Development and cosmeceutical evaluation of topical emulgel containing *Albizia lebbeck* bark extract. *J. Cosmet. Dermatol.* **2021**, *21*, 1588–1595. [CrossRef] [PubMed]

58. Togni, S.; Maramaldi, G.; Di Pierro, F.; Biondi, M. A cosmeceutical formulation based on boswellic acids for the treatment of erythematous eczema and psoriasis. *Clin. Cosmet. Investig. Dermatol.* **2014**, *7*, 321–327. [CrossRef] [PubMed]
59. Iram, F.; Khan, S.A.; Husain, A. Phytochemistry and potential therapeutic actions of Boswellic acids: A mini-review. *Asian Pac. J. Trop. Biomed.* **2017**, *7*, 513–523. [CrossRef]
60. Mun, S.; Kim, Y.L.; Kang, C.G.; Park, K.H.; Shim, J.Y.; Yong, P.K. Development of Reduced Fat Mayonnaise using 4αGT a semodified Rice Starch and Xanthan Gum. *Int. J. Biol. Macromol.* **2009**, *44*, 400–407. [CrossRef] [PubMed]
61. Yousuf, M.; Khan, H.M.S.; Rasool, F.; Khan, K.U.R.; Usman, F.; Ghalloo, B.A.; Umair, M.; Babalghith, A.O.; Kamran, M.; Aadil, R.M.; et al. Chemical Profiling, Formulation Development, In Vitro Evaluation and Molecular Docking of *Piper nigrum* Seeds Extract Loaded Emulgel for Anti-Aging. *Molecules* **2022**, *27*, 5990. [CrossRef]
62. Mehrzadi, S.; Tavakolifar, B.; Huseini, H.F.; Mosavat, S.H.; Heydari, M. The Effects of *Boswellia serrata* Gum Resin on the Blood Glucose and Lipid Profile of Diabetic Patients: A Double-Blind Randomized Placebo-Controlled Clinical Trial. *J. Evid.-Based Integr. Med.* **2018**, *23*, 2515690X18772728. [CrossRef] [PubMed]
63. Spiegel, M.; Cel, K.; Sroka, Z. The mechanistic insights into the role of pH and solvent on antiradical and prooxidant properties of polyphenols—Nine compounds case study. *Food Chem.* **2023**, *407*, 134677. [CrossRef] [PubMed]
64. Bengoechea, C.; Romero, A.; Aguilar, J.M.; Cordobés, F.; Guerrero, A. Temperature and pH as factors influencing droplet size distribution and linear viscoelasticity of O/W emulsions stabilized by soy and gluten proteins. *Food Hydrocoll.* **2010**, *24*, 783–791. [CrossRef]
65. Masuda, T.; Chinen, H.; Fukada, K. Effects of Protein Concentration and pH on Oil Droplet Size in O/W Emulsions Stabilized by Bovine Serum Albumin. *Chem. Lett.* **2014**, *43*, 598–600. [CrossRef]
66. Lukić, M.; Pantelić, I.; Savić, S.D. Towards Optimal pH of the Skin and Topical Formulations: From the Current State of the Art to Tailored Products. *Cosmetics* **2021**, *8*, 69. [CrossRef]

applied sciences

MDPI

Article

Chaves Thermal Spring Water Impact on Skin Health: Potential Cosmetic Application

Inês Pinto-Ribeiro [1], Cláudia Castro [1], Pedro Emanuel Rocha [1], Maria João Carvalho [1], Ana Pintado [1], Adélia Mendes [2], Sílvia Santos Pedrosa [1], Paula Capeto [1], João Azevedo-Silva [2], Ana L. S. Oliveira [1], Manuela Pintado [1] and Ana Raquel Madureira [1,*]

[1] CBQF—Centro de Biotecnologia e Química Fina, Laboratório Associado, Escola Superior de Biotecnologia, Universidade Católica Portuguesa, Rua Diogo Botelho 1327, 4169-005 Porto, Portugal; asoliveira@ucp.pt (A.L.S.O.)

[2] Amyris Bio Products Portugal, Unipessoal Lda, Rua Diogo Botelho 1327, 4169-005 Porto, Portugal; joao.pedro.silva@amyris.com (J.A.-S.)

* Correspondence: rmadureira@ucp.pt

Featured Application: Chaves thermal spring water, due to its potential anti-inflammatory and antipollution properties, has the potential to be incorporated into a cosmetic formulation.

Abstract: Since ancient times, thermal spring water has been proven to be beneficial to the skin and to improving dermatologic disorders, explaining its incorporation into cosmetic formulations as an active ingredient. Chaves thermal spring water, from northern Portugal, has been used as a local spa since Roman times, and its customers are satisfied with its medicinal quality. Despite the lack of published evidence on its specific effects on the skin, this study evaluates the potential of using Chaves thermal water as a cosmetic ingredient. The physiochemical composition demonstrated that Chaves thermal spring water is low-mineralized water, and its major components are sodium, potassium, silicon, and calcium. In vitro experiments demonstrated that this low mineralization might explain the absence of antioxidant and antiaging potential, and the maintenance of collagen and fibronectin levels. The quantification of the IL-6 levels showed that Chaves thermal spring water could be used as an anti-inflammatory product, suggesting its use by individuals with skin diseases. In agreement with this result, in vivo experiments revealed that Chaves thermal spring water improved the integrity of the skin barrier and preserved the skin microbial community. Overall, the present work suggests that Chaves thermal spring water might be used as a cosmetic product.

Keywords: thermal spring water; cosmetic application; anti-inflammatory properties

1. Introduction

Thermal water is naturally formed under specific geological conditions, and it is characterized as being rich in minerals and bacteriologically pure [1]. Depending on their geological, geochemical, and geothermic conditions, different springs can have very diverse chemical and physical water properties [2]. Since ancient times, thermal water has been associated with therapeutic effects; however, it is only more recently that some of these effects have been scientifically verified [3]. Bathing in thermal water has been proven to be beneficial to the skin and to improving dermatologic disorders such as atopic dermatitis [4,5] and psoriasis [6,7]. Such properties prompted the World Health Organization to officially recognize hydrotherapy and balneotherapy's efficiency within its Pain Management Protocol [8,9]. Given the growing emphasis on sustainable and natural remedies in modern society [10], there has been an observed and noticeable increase in the interest in and demand for the use of these types of therapies.

When undergoing balneotherapy, the first contact point between the human body with the water is the skin. As the largest human organ, the skin's primary function is to protect the body against the external environment, therefore making it the organ most impacted by the properties of thermal waters. The skin consists of three layers: the epidermis (the most exterior layer), the dermis (the middle layer), and finally, the hypodermis (subcutaneous tissue) [11]. The epidermis is a stratified epithelium composed of four layers of keratinocytes in progressive stages of differentiation [12]. The dermis is predominantly composed of fibroblast cells, but it also contains structures such as blood vessels, lymphatics, nerves, sweat glands, and pilosebaceous units. Lastly, the hypodermis is formed by subcutaneous fat and connective tissue [13]. Additionally, the skin supports a community composed of diverse bacteria, fungi, and viruses, which is known as the skin microbiota. In healthy individuals, the skin microbiota is in equilibrium with the host and is essential for skin health [14].

Exposure to pollution can cause oxidative stress, premature aging, inflammation, and diseases, not only in the respiratory and cardiovascular systems but also on the human skin [15]. Since most people's skin is exposed to pollution, protecting it is important [16]. Thermal water is rich in minerals that are assimilated by the skin and, depending on the mineral water composition and concentration, improve its physiological functions [8]. Sulfur-rich thermal waters have been proven to have antioxidant and anti-inflammatory effects [17–19] and potential antiaging properties for skin [20]. Thermal water also protects and repairs the skin barrier [21,22], being able to soothe sensitive skin [23]. Regarding the skin microbiota, there is evidence that thermal water increases its diversity and decreases inflammatory infectious agents [24,25].

The beneficial effects of thermal spring water on the skin and the skin microbiota have led to its incorporation into cosmetic formulations as an active ingredient [26]. In the present work, the potential of using Chaves thermal water as a cosmetic ingredient is investigated. Chaves thermal spring water is in the north of Portugal, and it has been used as a local spa since Roman times. Its waters are hot waters (ranging from 55 to 76 °C), with total diluted solids of 1600–1800 mg/L and a neutral pH, and they are bicarbonate-/sodium-/CO_2-rich [27]. Even though Chaves thermal spa has been functioning for several decades and its customers are satisfied with the medicinal quality of the thermal spring water [28], there is no published evidence, as far as we are aware, on the specific effects that this thermal spring water has on the skin. Therefore, the goal of this work was to assess the Chaves thermal spring water antioxidant, antiaging, and antipollution potential, and its impact on the skin biometric parameters and microbiota.

2. Materials and Methods

Thermal spring water was provided by Caldas de Chaves (Chaves, Portugal) and was directly retrieved at its source at 76 °C and bottled in 1-liter glass containers. It was then transported at room temperature, for approximately 1 h, and stored to 2–8 °C. Upon arrival, three different batches were filtered by vacuum with a 0.22 μm filter (Thermo Fisher Scientific, Waltham, MA, USA) to guarantee sterility.

2.1. Chaves Thermal Water Characterization

The pH value and electrical conductivity (EC) were determined using a multiparameter SevenExcellence^{TM} Mettler-Toledo AG (Greifensee, Switzerland) at 20 °C. The pH was measured with a precision of ±0.002 pH units, and the electrical conductivity was expressed as μS/cm.

The total dissolved solids (TDS) procedure was based on the standard method 2540C present in *Standard Methods for the Examination of Water and Wastewater*^{TM} [29]. Briefly, the method consists of filtering a well-mixed sample through a standard glass fiber filter, evaporating the filtrate to dryness in a weighed dish, and drying it to a constant weight at 180 °C. The increase in the dish weight represents the total dissolved solids.

For the mineral determination, 2 mL of Chaves thermal spring water was mixed with 5 mL of 65% HNO_3 plus 1 mL 30% H_2O_2 in a Teflon reaction vessel and digested in a microwave system (Speedwave, Berghof, Eningen, Germany). Digestion was conducted as follows: 160 °C for 5 min; 190 °C for 10 min; and 100 °C for 4 min. The resulting solutions were brought up to 10 mL with ultrapure water for analysis. Mineral concentrations were analyzed by inductively coupled plasma argon spectrometry (ICP; ICP-OES Optima 7000 DV, PerkinElmer, Waltham, MA, USA). The analyses were conducted in triplicate. Mineral concentrations were expressed in mg/L.

Microbial control was performed according to ISO 6222:1999 [30] "Water quality—Enumeration of culturable micro-organisms—Colony count by inoculation in a nutrient agar culture medium", for total germs. Briefly, 1 mL of Chaves thermal spring water was diluted in 9 mL of sterile peptone water and mixed in a vortex. Serial dilutions in peptone water were performed, and samples were plated by spread plate in tryptic soy agar (TSA) and Sabouraud dextrose agar (SDA). TSA plates were incubated at 37 °C for 24 h, while SDA plates were incubated at 30 °C for 48 h. The assay was performed in triplicate. After incubation, colony-forming units (CFUs) were calculated to determine the total aerobic bacteria and total yeasts and molds using the following formula:

$$\text{CFU} = \text{n}^{\circ}\text{colonies} \times \frac{1}{\text{v}} \times \frac{1}{\text{df}}$$

2.2. Evaluation of the Antioxidant Capacity of CHAVES Thermal Spring Water

The 2.2′-azino-bis (3-ethylbenzothiazoline-6-sulphonic acid) diammonium salt radical cation (ABTS) decolorization assay was performed as previously described [31]. In brief, 20 mL of ABTS solution was obtained by the addition of 7 mmol/L of ABTS (Sigma-Aldrich, St. Louis, MO, USA) to a 2.45 mM of potassium persulfate ($K_2S_2O_8$) solution (Merck, Darmstadt, Germany) at a 1:1 (*v*/*v*) proportion. The solution was left in the dark for 16 h and then diluted with deionized water to obtain an initial optical density (OD) of 0.700 ± 0.020 at 734 nm. An amount of 15 µL of the Chaves thermal spring water samples was placed in a 96-well plate with 200 µL of the diluted ABTS solution, in duplicate, and incubated for 5 min at 30 °C. A calibration curve with Trolox standard solutions (0.075–0.008 mg/mL) (Sigma-Aldrich, St. Louis, MO, USA) was included. After incubation, the OD was measured at 734 nm using a Synergy H1 microplate reader (Biotek, Winooski, VT, USA), and results were expressed as the percentage of inhibition of the free radicals generated.

Additionally, the 2,2-diphenyl-1-picrylhydrazyl (DPPH) salt radical cation decolorization assay was performed as already reported [31]. Firstly, a 100 mL solution of 600 µM DPPH was prepared combining 24 mg of DPPH salt (Sigma-Aldrich, St. Louis, MO, USA) with methanol. This solution was then diluted with methanol to obtain a DPPH solution with an OD of 0.600 ± 0.100 at 515 nm. Afterwards, 25 µL of the Chaves thermal spring water was added (in duplicate) to a 96-well plate with 175 µL of the DPPH solution. It also included a calibration curve prepared with Trolox standard solutions (0.075–0.008 mg/mL). The plate was incubated at room temperature for 30 min, followed by the measurement of the OD at 515 nm. The results were expressed as the percentage of inhibition of the free radicals generated.

2.3. Skin Enzyme Inhibition

2.3.1. Elastase Inhibition Assay

The Neutrophil Elastase Inhibitory assay kit (Abcam, Cambridge, UK) was performed to evaluate the elastase inhibitory capacity of Chaves thermal spring water. In a 96-well plate, a mixture of 25 µL of Chaves thermal spring water and 50 µL of elastase solution was incubated for 5 min at 37 °C. Then, the substrate solution was added to each well, and the fluorescence was measured at Ex/Em = 400/505 nm for 30 min at 37 °C using Synergy H1 (Biotek, Vila Nova de Gaia, Portugal). Each assay included the blank, 5.0×10^{-4} mg/mL of peptide succinyl-alanyl-alanyl-prolyl-valine chloromethyl ketone (SPCK) as a positive control of inhibition, and 8 mg/mL of vitamin C (benchmark). The RFU of the fluorescence produced by substrate hydrolysis was determined by ΔRFU = R2 − R1. It is advised to read kinetically and select R1 and R2 in the linear range. The percentage inhibition was calculated using the following equation:

$$\text{Enzyme inhibition activity } (\%) = \frac{\Delta\text{RFU test inhibitor}}{\Delta\text{RFU enzyme control}} \times 100$$

2.3.2. Collagenase Inhibition Assay

The colorimetric metalloproteinases 1 (MMP1) inhibitor screening kit (abcam, Cambridge, UK) was performed according to the manufacturer's instructions. Briefly, the 0.75 U/L MMP1 enzyme was introduced to a flat-bottom 96-well microplate (Thermo Scientific, MA, USA), followed by the addition of Chaves thermal spring water samples, vitamin C (benchmark), N-Isobutyl-N-(4-methoxyphenylsulfonyl)glycyl hydroxamic acid (NNGH) as a positive control, and the blank to the respective wells. Then, the microplate was incubated at 37 °C for 30 to 60 min, facilitating the interaction between the test samples or controls and the enzyme. Finally, 1 mM thiopeptide chromogenic substrate (Ac-PLG-[2-mercapto-4-methyl-pentanoyl]-LG-OC_2H_5) was added, and the absorbance was recorded at 412 nm using Synergy H1 (Biotek, Vila Nova de Gaia, Portugal). The data were collected for 10 to 20 min at intervals of 1 min. The percentage of enzyme inhibition activity was calculated using the following equation, in which V is the reaction velocity expressed in OD/min:

$$\text{Enzyme inhibition activity } (\%) = 1 - \left(\frac{\text{V inhibitor}}{\text{V enzyme control}} \times 100\right)$$

2.3.3. Tyrosinase Inhibition Assay

A colorimetric tyrosinase inhibitor assay kit (abcam, Cambridge, UK) was used. Shortly, a 96-well plate containing a mixture of 20 µL of Chaves thermal spring water samples and 50 µL tyrosinase solution was incubated at 25 °C for 10 min. Then, the substrate solution was added, and the absorbance was measured at 510 nm using Synergy H1 (Biotek, Vila Nova de Gaia, Portugal). Throughout a 30 min period, the data were monitored at 2 min intervals. Each assay included a blank, 0.021 mg/mL of Kojic acid as a positive control of inhibition, and 8 mg/mL vitamin C (benchmark). To determine the equivalent values for absorbance (A1 and A2), two time points (T1 and T2) were selected in the linear range of the plot. The slopes were computed for each sample (S), inhibition control (IC), and enzyme control (EC) by dividing the ΔA (A2-A1) values by the time (ΔT) (T2-T1). The percentage of enzyme inhibition activity was calculated using the following equation:

$$\text{Enzyme inhibition activity } (\%) = \frac{\text{Slope of EC} - \text{Slope of S}}{\text{Slope of EC}} \times 100$$

2.4. Cell Culture Assays

Different types of cells were used, including the Human Keratinocyte (HaCaT) cell line (CLS, Lot No. 300493-4619) and Human Primary Dermal Fibroblasts (nHDFs) from adult skin (Lonza Bioscience, Cat. CC2511, Lot No. 0000577924). Both cell lines were cultured in Gibco Dulbecco's Modified Eagle Medium (DMEM) supplemented with FBS (Gibco, Thermo Fisher Scientific, USA) plus penicillin (100 U/mL)–streptomycin (100 µg/mL) (Gibco, Thermo Fisher Scientific, USA) and maintained at 37 °C in a 5% CO_2 humidified atmosphere.

For in vitro assays, two additional media were prepared: DMEM powder high-glucose medium (Gibco, Thermo Fisher Scientific, USA) with 200 mL of Chaves thermal spring water (test culture medium) and DMEM powder high-glucose medium with 200 mL of ultrapure Mili-Q-type water (control culture medium). The final pH values of the culture media were adjusted to 7.4. Finally, culture media were supplemented with FBS plus penicillin (100 U/mL)–streptomycin (100 µg/mL).

2.4.1. Cytotoxicity Assay

HaCaT and nHDF lines were seeded at 1 × 10^4 cells/well in a 96-well plate and left to adhere overnight at 37 °C in a 5% CO_2 humidified atmosphere. Afterwards, culture medium was replaced by test culture- or control culture-supplemented media and cells were re-incubated for 24 h at 37 °C in a 5% CO_2 humidified atmosphere. To evaluate the cytotoxicity, 10% (*v*/*v*) of 10× PrestoBlue cell viability reagent (Invitrogen, Thermo Fischer Scientific, Waltham, MA, USA) was added to each well and incubated for 1 h at 37 °C in 5% CO_2, protected from direct light. Then, the cell viability was detected by fluorescence using Synergy H1 (Biotek, Vila Nova de Gaia, Portugal), and morphological cell alterations were monitored by microscope observation. The experiments were performed in triplicates using HaCaT cells on the 20th and 22nd passages and nHDF cells on the 5th and 6th passages. The results were presented as the percentage of cell viability, where 100% viability corresponds to control cells and a reduction in the cell viability of more than 30% is considered a cytotoxic effect, according to ISO 10993-5 [32].

2.4.2. Quantification of Pro-Collagen 1 α1 and Fibronectin

nHDFs were seeded in 12-well plates at 3 × 10^5 cells/well and allowed to adhere overnight at 37 °C in a 5% CO_2 humidified atmosphere. Next, the culture medium was replaced by test culture- or control culture-supplemented media and cells were incubated for 24 h at 37 °C in a 5% CO_2 humidified atmosphere. At the end, the total protein was obtained using the lysis buffer of the ELISA kits, and its concentration was determined by the BCA kit (Thermo Fischer, Waltham, MA, USA). The amount of total protein of all the samples was normalized to 10 μg/mL. The pro-collagen 1 α1 was quantified by the Human Pro-Collagen I alpha 1 ELISA Kit (abcam, Cambridge, UK), while the fibronectin quantification was performed using the Human Fibronectin ELISA Kit (abcam, Cambridge, UK), according to the manufacturer's instructions. An amount of 0.5 μM Palmitoyl Tripeptide-1 (Pal-GHK, Cayman Chemicals, Ann Arbor, MI, USA) was used as a positive control for both assays, which were performed in triplicate.

2.4.3. Exposure to Urban Particulate Matter

HaCaT cells were seeded at 1 × 10^5 cells/well in a 24-well plate and maintained at 37 °C in a 5% CO_2 humidified atmosphere for 24 h. Afterwards, cells were incubated with and without 500 μg/mL of urban air pollution particles (SRM 1648a) resuspended in test- or control-supplemented medium for 24 h, following the literature [33]. The supernatants were collected and used to evaluate the levels of proinflammatory cytokines, such as IL-6, by ELISA (Biolegend, San Diego, CA, USA). Cells were lysed with water and used for protein quantification via the BCA method (Thermo Fischer, Waltham, MA, USA). Two independent experiments were performed, and the results were expressed in pg of cytokine/mg of total protein.

2.5. Population of the Study of Skin Microbiota

For the present study, female and male volunteers more than 18 years old with or without skin diseases were recruited. The group of selected volunteers did not include pregnant women or women during the lactation period; individuals with tattoos or significant scars on the inner forearm; individuals that performed hair removal/exfoliation/skin cleansing on the inner forearm 3 to 4 weeks prior to the sampling and during the study period. We also excluded individuals that applied cosmetic products onto the inner forearm prior to the sampling and during the study, and those who took pre- or probiotics, antibiotics, immunosuppressants, and chronic anti-inflammatory and chronic antihistamine drugs and/or systemic antifungals 1 month prior to the sampling and during the study period. Based on these exclusion criteria, a total of 23 participants were included (19 females and 4 males), who were divided into two age groups: 26 to 35 years old (n= 17) and 36 to 45 years old (n = 6). Seven of the volunteers mentioned eczema, neurofibromatosis, psoriasis, and sensitive skin. Oral and written instructions were provided to all volunteers, who received a 50 mL vaporizer with Chaves thermal spring water to be used twice a day (firstly in the morning and lastly in the afternoon) for 15 days.

Volunteers visited the locale of the study on days 0, 8, and 16, on which the measurement of the skin biometric parameters of both inner forearms was performed. Additionally, samples of skin microbiota were collected from both inner forearms of 13 of the total selected volunteers. The control and test forearms were randomly chosen to eliminate the effect of the dominant arm. All volunteers signed an informed consent form after receiving a detailed explanation about the purpose and procedures of the study. Skin microbial samples and skin biometric data were delinked and unidentified from their donors. The present study (project no. 83) was validated by the Health Ethics Committee of the Portuguese Catholic University (CES-UCP) before its execution.

2.6. Measurement of Skin Biometric Parameters

To evaluate the beneficial effects of Chaves thermal spring water on the skin, the biometric parameters were measured on the inner forearm before (day 0) and after (days 8 and 16) the application of the product. Twelve hours before the measurement, the volunteers were instructed not to apply any cosmetic or the test product. A Multi Probe Adapter MPA 6 (Courage and Khazaka, Cologne, Germany) coupled with different probes was used: the Corneometer® CM 825 probe, to quantify the level of hydration; the Tewameter® TM Hex probe, to determine the transepidermal water loss (TEWL); and the probe Skin-pH-meter PH 905, to measure the pH.

2.7. Collection of Skin Microbiota

On days 0 and 16, samples of skin microbiota from the inner forearms of 13 volunteers were collected using the procedure previously described by our team [34]. Briefly, 4N6FLOQSwabs™ (Thermo Fisher Scientific, Waltham, MA, USA) moistened in a sterile solution of phosphate buffer solution (PBS) (at 0.1 M, pH 7.3 ± 0.2 at 25 °C) plus 0.1% (v/v) Tween 80 were used to collect the skin microbiota samples from each inner forearm. After the collection, the swab was placed into a tube with RPMI 1640 medium (Gibco, Thermo Fisher Scientific, Waltham, MA, USA), and it was incubated for 2 h at 34 °C with agitation. The control of the collection method was performed using the same procedure without the skin microbiota sample. At the end, all tubes were centrifuged at 21,130× g for 10 min, and the pellet was recovered and stored at −20 °C until DNA extraction. The controls were processed similarly to the skin microbiota samples.

2.8. DNA Extraction and Quantitative Real-Time PCR (qPCR)

Total DNA was extracted from all pellets with the QIAamp DNA Microbiome Kit (Qiagen, Hilden, Germany), according to the manufacturing instructions. After extraction, DNA was quantified by the Qubit 4 Fluorometer dsDNA HS Assay Kit (Life Technologies, Foster City, CA, USA), and the concentration was standardized at 10 ng/µL.

qPCR assays were used to determine the relative abundances of specific microbial genera and species. For this, we used a universal assay, composed by universal primers targeting a conserved region of the 16S rRNA gene for bacteria and the ITS2 region for fungi, and genus- or species-specific assays, composed by primers targeting genus- or species-specific genes. qPCR reactions were prepared as previously described by our team [34].

2.9. Statistical Analysis

Data were plotted and treated using GraphPad Prism version 6.00 (GraphPad Software, Insight Partners, La Jolla, CA, USA). Data were further analyzed for significant differences. Multiple comparison tests were performed by one-way ANOVA supplemented with Tukey's HSD post hoc test. Differences were considered statistically significant at $p < 0.05$.

3. Results

3.1. Characterization of Chaves Thermal Water

Chaves thermal spring water is characterized by a neutral pH (6.8) and low mineralization (1630 mg/L of total dissolved solids), and its main minerals are sodium, potassium, silicon, and calcium (Table 1). Additionally, no microbial growth was detected on the TSA and SDA plates.

Table 1. Elemental composition of Chaves thermal spring water (recalculated) and physicochemical properties.

Chaves Thermal Spring Water	
Physicochemical characterization	
pH	6.84 ± 0.01
Conductivity (μS/cm)	2532 ± 2.12
Total dissolved solids (mg/L)	1630 ± 10
Minerals (mg/L) *	
Sodium (Na)	576 ± 16.91
Potassium (K)	74.02 ± 1.02
Silicon (Si)	37.20 ± 0.34
Calcium (Ca)	26.03 ± 0.46
Magnesium (Mg)	7.62 ± 0.10
Sulfur (S)	4.89 ± 0.05
Phosphor (P)	4.02 ± 0.05
Molybdenum (Mo)	0.04 ± 0.001
Manganese (Mn)	0.03 ± 0.0004
Cadmium (Cd)	0.02 ± 0.0003

* The elemental concentrations were recalculated from the measured concentrations of chemical species present in the water to represent the elemental equivalents.

3.2. Cytotoxicity Assessment

The impact of Chaves thermal spring water on the viability of nHDF and HaCaT cells was evaluated by the PrestoBlue cell viability assay. As represented in Figure 1, the exposure to Chaves thermal spring water did not significantly reduce the nHDF or HaCaT cell viability when compared to the control. According to ISO 10993-5, an ingredient is considered cytocompatible when the cell viability inhibition is lower than 30%.

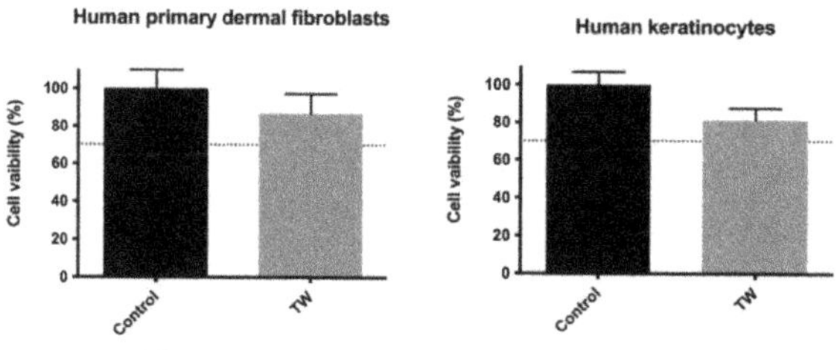

(a) (b)

Figure 1. Effect of Chaves thermal spring water on cellular viability in (**a**) human dermal fibroblasts (nHDFs) and (**b**) keratinocytes (HaCaT) after 24 h incubation. nHDF and HaCaT cells were incubated with DMEM powder high-glucose medium plus Chaves thermal spring water (TW) and with DMEM powder high-glucose plus Mili-Q-type water (control). The pH values of both media were adjusted to 7.4, and the media were supplemented with FBS plus penicillin–streptomycin. Results are presented as percentage of cell viability, where 100% corresponds to untreated cells. The dotted line represents a 30% inhibition of cell viability. Statistical analysis was performed using the one-way ANOVA with Tukey's multiple comparison test.

3.3. Antioxidant and Antiaging Enzyme Activities

To investigate the antiaging potential of Chaves thermal spring water, commercial kits were used to determine the percentage of inhibition of aging-related enzymes, including elastase, collagenase, and tyrosinase (Table 2). Chaves thermal spring water significantly inhibited the activity of elastase in comparison to vitamin C, while the activity of MMP-1 (collagenase) and tyrosinase was not reduced. Vitamin C (at 8 mg/mL) showed an inhibition capacity to all the tested enzymes between 90 and 99%.

Table 2. Evaluation of antioxidant and antiaging potential of Chaves thermal spring water.

	Enzyme Relative Inhibition (%)			Antioxidant Activity (%)	
	Elastase	MMP-1	Tyrosinase	ABTS	DPPH
Chaves thermal spring water	14.22 ± 4.29 [b]	0.70 ± 0.25 [b]	0.005 ± 0.00 [b]	9.09 ± 3.88 [b]	ND
Vitamin C (8 mg/mL)	90.54 ±1.51 [a]	96.26 ± 0.31 [a]	99.26 ± 0.31 [a]	–	–
Vitamin C (0.075 mg/mL)	–	–	–	91.94 ± 2.03 [b]	86.56 ± 0.48 [b]

ND—not detected. Elastase positive control used was SPCK (5.0×10^{-4} mg/mL), for MMP-1 or collagenase, NNGH (4.1×10^{-5} mg/mL) was used, and for tyrosinase, kojic acid was used (0.021 mg/mL). [a,b] Different letters represent the significant differences ($p < 0.05$) between Chaves thermal spring water and vitamin C.

Additionally, two chemical methods (ABTS and DPPH) were used to assess the antioxidant activity of Chaves thermal spring water. However, only the ABTS method allowed for an evaluation of the neutralization of free radicals by Chaves thermal spring water (Table 2).

3.4. Cosmetic and Skincare Properties

3.4.1. Collagen and Fibronectin Production

The Human Pro-Collagen I alpha 1 ELISA Kit demonstrated a similar quantity of pro-collagen I alpha 1 protein between nHDF cells incubated with the test-supplemented medium and those incubated with the control-supplemented medium (Figure 2a). In agreement with this result, the quantity of fibronectin protein detected in nHDF cells incubated with the test-supplemented medium did not show statistical differences when compared with nHDF cells incubated with the control-supplemented medium (Figure 2b).

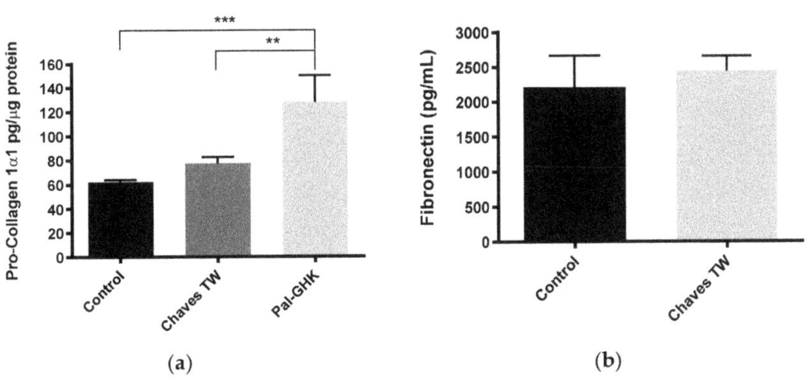

Figure 2. Quantification of (**a**) pro-collagen 1 α1 and (**b**) fibronectin in nHDF cells treated with Chaves thermal spring water (Chaves TW). The control condition of both assays corresponds to standard culture medium prepared with Mili-Q water. Palmitoyl Tripeptide-1 (Pal-GHK) was used as a positive control. Statistical analysis was performed using the one-way ANOVA with Tukey's multiple comparison test (** $p < 0.001$, *** $p < 0.0001$).

3.4.2. Cellular Exposure to Urban Particulate Matter

To evaluate the anti-inflammatory potential of Chaves thermal spring water, HaCaT cells were exposed to urban air pollution particles, followed by IL-6 quantification, which is a sensitive and reliable inflammation marker due to its involvement in the acute-phase response [35]. It was possible to observe that in the basal conditions (without urban air particles), Chaves thermal spring water led to a significant decrease in the IL-6 levels in comparison to the control condition (Figure 3a). Moreover, the levels of IL-6 increased 10-fold after the exposure to urban air pollution particles. The treatment with Chaves thermal spring water led to a significant reduction in the IL-6 levels in comparison to the control condition. The betamethasone (potent anti-inflammatory molecule) was used as a positive control (Figure 3b).

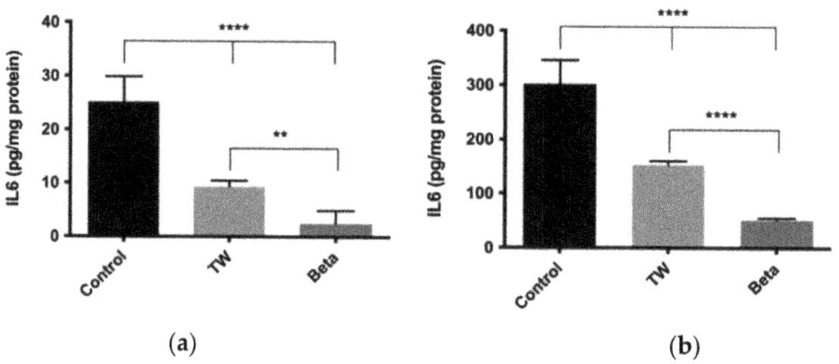

Figure 3. Evaluation of anti-inflammatory potential of Chaves thermal water. The IL-6 levels were quantified in supernatants of HaCaT cells (**a**) without urban air pollution particles and (**b**) upon contact with urban air pollution particles. HaCaT cells were incubated with DMEM powder high-glucose medium plus Chaves thermal spring water (TW) and with DMEM powder high-glucose plus Mili-Q-type water (control). The pH values of both media were adjusted to 7.4, and the media were supplemented with FBS plus penicillin–streptomycin. Betamethasone (Beta) was used as a positive control for the anti-inflammatory effect. (** $p < 0.001$, **** $p < 0.0001$).

3.5. Evaluation of Skin Parameters and Skin Microbiota

The skin biometric parameters, including the hydration levels, transepidermal water loss (TEWL), and pH values, were evaluated before (day 0) and after (days 8 and 16) the application of Chaves thermal spring water (Figure 4). The hydration levels of both inner forearms of 23 volunteers were similar when comparing day 0 with day 8 and day 16 (Figure 4a). To evaluate whether the hydration levels were influenced by gender, age, or the presence/absence of skin diseases, the results were analyzed based on these criteria, which demonstrated that the hydration levels were maintained during the time. Interestingly, when the volunteers were grouped based on their hydration levels at day 0 (before the application of Chaves thermal spring water), the group of volunteers with very dry skin (hydration levels < 30) revealed a statistically significant increase in their hydration levels at day 8, maintaining this until day 16.

On day 16, our results demonstrated that the application of Chaves thermal spring water significantly decreased the transepidermal water loss to values under 10 $g/m^2/h$ in comparison to day 0, meaning that this thermal water might help to restore the integrity of the skin barrier (Figure 4b). Additionally, the skin pH values of the volunteers increased significantly on days 8 and 16 in comparison to day 0. However, the volunteers presented similar skin pH values on days 8 and 16, demonstrating that Chaves thermal spring water allowed them to maintain a healthy skin pH (Figure 4c).

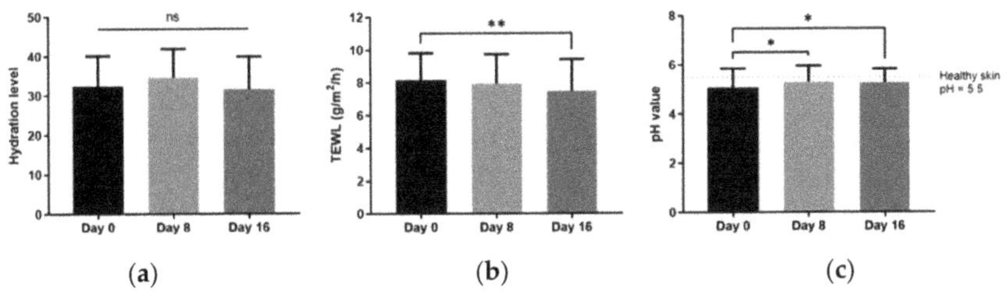

Figure 4. Skin biometric parameters. The measurement of (**a**) hydration levels, (**b**) transepidermal water loss (TEWL), and (**c**) pH values was performed on days 0 (before the application) and on days 8 and 16 (after the application). The results are represented as bar graphs (average ± SD). * and ** stand for $p < 0.05$ and $p < 0.005$, respectively. ns, not significant ($p > 0.05$).

To further evaluate the impact of Chaves thermal spring water on skin health, we also characterized the skin microbiota in a subset of 13 volunteers at days 0 and 16. *Staphylococcus, Propionibacterium,* and *Corynebacterium* are the main bacterial genera that compose the human skin microbiota [36]. Thus, their relative abundances were determined on the samples collected from the skin of the inner forearms of volunteers on day 0 (before the application) and day 16 (after the application), which presented similar amounts of these genera (Figure 5a–c). We also determined the relative abundances of *S. epidermidis* and *P. acnes* (skin health sentinels), demonstrating no statistically significant differences between the skin microbiota samples collected before and after the application of Chaves thermal spring water (Figure 5e,f). A similar result was obtained for the *Staphylococcus* sp./*Propioniobacterium* sp. and *S. epidermidis*/*P. acnes* ratios when comparing both time points of collection. Moreover, the comparison of samples collected on day 0 and those collected on day 16 showed that the relative abundance of *Malassezia* genus was similarly detected at both time points (Figure 5d).

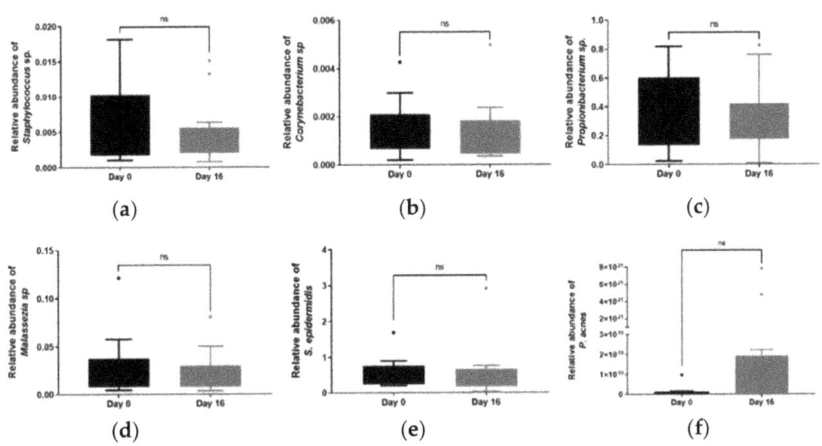

Figure 5. The relative abundances of (**a**) *Staphylococcus* sp.; (**b**) *Corynebacterium* sp.; (**c**) *Propionibacterium* sp.; (**d**) *Malassezia* sp.; (**e**) *Staphylococcus epidermidis;* (**f**) *Propionibacterium acnes,* determined by qPCR (ns, not significant).

4. Discussion

For centuries, thermal spring water has been used as a treatment for several diseases, including dermatologic diseases [3]. Different studies have reported the absence of side effects of thermal spring water, allowing for its use as an adjuvant in the treatment of various skin disorders, demonstrating the importance of thermal spring waters in the cosmetics industry [26]. The scientific knowledge about the mechanisms of action and clinical benefits of this natural resource has been increasing over the years, including in Portugal [23,37–40]. In the present study, we performed a preliminary screening of Chaves thermal spring water, evaluating its biological potential for use as a therapeutic or cosmetic product.

We began by determining the physicochemical composition of Chaves thermal spring water, which was consistent with previous reports [27], indicating that Chaves thermal spring water has been stable over the years. Its pH value was 6.8, an almost neutral value. This is an important characteristic for a potential cosmetic active ingredient that will be topically applied, since skin pH is normally acidic, ranging between 4 and 6, which regulates the maintenance of the stratum corneum homeostasis and the skin barrier permeability [41]. Therefore, skincare products should have a pH similar to that of skin so they do not alter it [42]. Additionally, Chaves thermal spring water contained less than 2 g/L of total dissolved solids and is classified as low-mineralized in comparison to other thermal waters, but this does not impair its potential to benefit the skin, since the absorption of minerals through the skin is limited [43]. For example, Avène thermal spring water is also low-mineralized, and it presents anti-inflammatory, antioxidant, and anti-irritant proprieties [23,44,45]. The main minerals presented in Chaves thermal spring water, sodium, potassium, silicon, and calcium, are also present in other thermal spring waters around the globe, but in different proportions [26]. After the determination of the physicochemical composition, a cytotoxicity assay was performed, demonstrating that Chaves thermal spring water did not interfere with the viability of HaCaT and nHDF cells, allowing it to go through further in vitro tests. The non-toxicity of other thermal spring waters has also been reported by other authors [46–48].

Recently, Figueiredo et al. [26] reviewed the published data on the dermatologic potential of thermal spring waters, based on our evaluation of the antiaging and antioxidant capacity of Chaves thermal spring water. Our results demonstrated that Chaves thermal spring water inhibited the activity of elastase, which is an enzyme that hydrolyses elastin, and its activity is associated with the mechanical properties of connective tissues, including the firmness and elasticity of the skin [49]. This result suggests that Chaves thermal spring water may contribute to the improvement in skin health while avoiding the degradation of the elastase.

The antioxidant activity of thermal spring waters is related to their mineral composition. In fact, minerals like calcium, potassium, and magnesium are involved in human cell defense mechanisms against reactive oxidative species (ROS), since they support the activity of several antioxidant enzymes. It is also known that mineral salts such as selenium, copper, manganese, and zinc have an important role as antioxidants, with selenium being used in the effective removal of peroxides from cytosol and cell membranes, while the cytosolic superoxide dismutase requires copper to activate its antioxidant properties [50]. For example, Avène, Uriage, and Vichy thermal spring waters, as well as dead sea water, present high contents of some of these minerals, and in vitro experiments have demonstrated their antioxidant capacity by reducing the generation of ROS in UV-exposed skin or cells [26,51,52]. In contrast, Chaves thermal spring water did not show antioxidant activity in comparison to vitamin C via the ABTS and DPPH methods. This result might be explained by its mineral composition, in which sodium is dominant.

Previous studies have reported that thermal spring waters have a positive effect on extracellular matrix (ECM) proteins, including collagen and fibronectin, as its major components. Collagen is responsible for skin elasticity and flexibility, while fibronectin plays an important role in the ECM organization and stability [26,51,52]. For instance, in vitro studies have demonstrated that Blue Lagoon and Nitrodi's spring waters significantly increased the levels of collagen 1a1 and 1a2 in human epidermal keratinocytes and fibroblasts [53,54]. It has been also reported that the silicon content of thermal spring water might be related to collagen synthesis via the activation of hydroxylation enzymes that are crucial for forming the collagen network, thereby improving skin elasticity and strength [55]. Although

Chaves thermal spring water contains 37.20 mg/L of silicon, this concentration was not sufficient to increase the collagen synthesis in nHDF cells. Moreover, in vitro studies have demonstrated a promotion in the fibronectin deposition with Nitrodi's spring water [53]. In contrast, our results demonstrated no significant increase in the production of fibronectin after the incubation of nHDF cells with Chaves thermal spring water.

The human skin acts as the exterior interface of the human body with the environment, meaning that the exposure to air pollution or UV radiation might have a negative impact on skin health, leading to inflammatory processes and ROS production. Several changes have been reported to occur in skin cells after their exposure to pollution factors, such as in lipid composition and protein oxidation, and increases in inflammation markers and oxidative stress [16]. In the present study, Chaves thermal spring water significantly decreased the levels of IL-6 in HaCaT cells exposed to urban air pollution particles. This result indicates the soothing potential of Chaves thermal spring water, which is in accordance with the published data for other thermal spring waters [56].

In addition to in vitro experiments, we measured the skin biometric parameters on the forearms of volunteers, demonstrating that Chaves thermal spring water significantly decreased the TEWL without changing the hydration level (comparing day 0 and day 16). The values of the TEWL and hydration level are associated with the health of the skin as well as the age of individuals, and sometimes they are inversely related [57]. Thus, both skin parameters are frequently used in dermatology and cosmetology to evaluate the integrity of the skin barrier and to assess the efficacy of topical products, such as thermal spring waters [40,57]. Our results suggested that Chaves thermal water might be used to restore the skin barrier function due to the decrease in the TEWL, making it helpful for individuals with skin diseases, who are associated with higher values of TEWL in comparison to the unaffected [58,59]. Montero-Vilchez et al. [58] reported that the TEWL was significantly higher in psoriatic plaques and atopic dermatitis eczematous lesions than in uninvolved or healthy skin. Different studies have demonstrated that thermal spring water might have prebiotic and probiotic effects based on their microbial and mineral compositions [60–63]. Chaves thermal spring water is a low-mineral water, mainly composed of sodium and potassium, with a stable microbial community over time [64]. These factors might explain the absence of differences between the skin microbiota samples collected before application (day 0) and those collected after thermal water application (day 16). In contrast, a study with 57 individuals with psoriasis showed that a 12-bath treatment at Terme di Comano (Trentino, Italy) helped to restore the microbial community of psoriatic lesions so that it resembled that of unaffected or peri-lesion skin [63]. Moreover, the 3-week balneotherapy treatment with La Roche Posay thermal water (selenium-rich water) significantly increased the level of the *Xanthomonas* genus, which was associated with a clinical improvement in psoriasis vulgaris [62].

5. Conclusions

Overall, the present study provides a comprehensive evaluation of the biological potential of Chaves thermal spring water as a cosmetic product.

Due to its closeness to a neutral pH level, Chaves thermal spring water reveals itself as an ideal candidate for topical application, without disrupting the skin's natural acidity. It also exhibits a low-mineralized composition, aligning it with other renowned, commercially available thermal waters known for their dermatological benefits.

The most notable characteristic of this water is its anti-inflammatory properties. As demonstrated here, Chaves thermal spring water significantly reduced the IL-6 levels in HaCaT cells that were exposed to urban pollution, making it relevant in the context of increasing environmental stressors affecting skin health.

Additionally, the clinical study here employed revealed a significant reduction in transepidermal water loss in the human volunteers, suggesting that Chaves thermal spring water may help restore the skin barrier function, which is a critical finding for individuals with skin conditions associated with an impaired barrier function.

In summary, this study establishes a solid foundation for the use of Chaves thermal spring water as a cosmetic ingredient, highlighting its benefits for skin health. However, it is necessary to perform additional studies, including those with a high number of volunteers with different skin diseases (such as psoriasis and eczema), to evaluate the beneficial effects of Chaves thermal water on affected skin.

Author Contributions: Conceptualization, A.R.M.; methodology, I.P.-R., C.C., P.E.R., M.J.C., A.R.M., S.S.P., P.C., J.A.-S. and A.L.S.O.; software, I.P.-R., S.S.P., J.A.-S. and A.P.; validation, A.R.M. and M.P.; formal analysis, I.P.-R., C.C., S.S.P., J.A.-S. and A.L.S.O.; investigation, I.P.-R., C.C., P.E.R., M.J.C., A.M., S.S.P., P.C., J.A.-S. and A.L.S.O.; resources, M.P. and A.R.M.; data curation, M.P. and A.R.M.; writing—original draft preparation, I.P.-R., P.E.R. and C.C.; writing—review and editing, P.E.R.; visualization, I.P.-R.; supervision, A.R.M. and M.P.; project administration, M.P. and A.R.M.; funding acquisition, A.R.M. and M.P. All authors have read and agreed to the published version of the manuscript.

Funding: This research was funded by the Project PL23-00053, "Therm4Skin—Bem-Estar Sem Pausa", Promove o futuro do Interior, call 2023 and CBQF under the FCT project UIDB/50016/2020 through national funds, and the co-funding by FEDER, within the PT02020 Partnership Agreement.

Institutional Review Board Statement: This study was conducted in accordance with the Declaration of Helsinki, and it was approved by the Health Ethics Committee of the Portuguese Catholic University (CES-UCP) (Project 83, approved in 17 September 2020) for studies involving volunteers.

Informed Consent Statement: Informed consent was obtained from all subjects involved in this study.

Data Availability Statement: The raw data supporting the conclusions of this article will be made available by the authors on request.

Acknowledgments: The authors thank the scientific collaboration of Amyris Bio Products Portugal.

Conflicts of Interest: Authors Adélia Mendes and João Azevedo-Silva were employed by the company Amyris Bio Products Portugal, Unipessoal Lda. The remaining authors declare that the research was conducted in the absence of any commercial or financial relationships that could be construed as a potential conflict of interest.

References

1. Ghersetich, I.; Freedman, D.; Lotti, T. *Balneology Today*; Wiley Online Library: Hoboken, NJ, USA, 2000; pp. 346–348.
2. Komatina, M. *Medical Geology: Effects of Geological Environments on Human Health*; Elsevier: Amsterdam, The Netherlands, 2004.
3. Huang, A.; Seité, S.; Adar, T. The use of balneotherapy in dermatology. *Clin. Dermatol.* **2018**, *36*, 363–368. [CrossRef] [PubMed]
4. Guerrero, D.; Garrigue, E. Eau thermale d'Avène et dermatite atopique: Avène's thermal water and atopic dermatitis. In *Annales de Dermatologie et de Vénéréologie*; Elsevier: Amsterdam, The Netherlands, 2017.
5. Geat, D.; Giovannini, M.; Barlocco, E.G.; Pertile, R.; Farina, S.; Pace, M.; Filippeschi, C.; Girolomoni, G.; Cristofolini, M.; Baldo, E. Characteristics associated with clinical response to Comano thermal spring water balneotherapy in pediatric patients with atopic dermatitis. *Ital. J. Pediatr.* **2021**, *47*, 91. [CrossRef] [PubMed]
6. Kulisch, Á.; Mándó, Z.; Sándor, E.; Lengyel, Z.; Illés, A.; Kósa, J.; Árvai, K.; Lakatos, P.; Tóbiás, B.; Papp, M.; et al. Evaluation of the effects of Lake Hévíz sulfur thermal water on skin microbiome in plaque psoriasis: An open label, pilot study. *Int. J. Biometeorol.* **2023**, *67*, 661–673. [CrossRef] [PubMed]
7. Khalilzadeh, S.; Shirbeigi, L.; Naghizadeh, A.; Mehriardestani, M.; Shamohammadi, S.; Tabarrai, M. Use of mineral waters in the treatment of psoriasis: Perspectives of Persian and conventional medicine. *Dermatol. Ther.* **2019**, *32*, e12969. [CrossRef] [PubMed]
8. Cacciapuoti, S.; Luciano, M.A.; Megna, M.; Annunziata, M.C.; Napolitano, M.; Patruno, C.; Scala, E.; Colicchio, R.; Pagliuca, C.; Salvatore, P.; et al. The role of thermal water in chronic skin diseases management: A review of the literature. *J. Clin. Med.* **2020**, *9*, 3047. [CrossRef]
9. Antunes, J.D.M.; Daher, D.V.; Giaretta, V.M.D.A.; Ferrari, M.F.M.; Posso, M.B.S. Hydrotherapy and crenotherapy in the treatment of pain: Integrative review. *BrJP* **2019**, *2*, 187–198. [CrossRef]
10. Meier, B.P.; Dillard, A.J.; Osorio, E.; Lappas, C.M. A behavioral confirmation and reduction of the natural versus synthetic drug bias. *Med. Decis. Mak.* **2019**, *39*, 360–370. [CrossRef]
11. Hwa, C.; Bauer, E.A.; Cohen, D.E. Skin biology. *Dermatol. Ther.* **2011**, *24*, 464–470. [CrossRef]
12. Lefèvre-Utile, A.; Braun, C.; Haftek, M.; Aubin, F. Five functional aspects of the epidermal barrier. *Int. J. Mol. Sci.* **2021**, *22*, 11676. [CrossRef]
13. Chambers, E.S.; Vukmanovic-Stejic, M. Skin barrier immunity and ageing. *Immunology* **2020**, *160*, 116–125. [CrossRef]
14. Carvalho, M.J.; Oliveira, A.L.S.; Pedrosa, S.S.; Pintado, M.; Pinto-Ribeiro, I.; Madureira, A.R. Skin microbiota and the cosmetic industry. *Microb. Ecol.* **2023**, *86*, 86–96. [CrossRef] [PubMed]

15. Schikowski, T.; Hüls, A. Air pollution and skin aging. *Curr. Environ. Health Rep.* **2020**, *7*, 58–64. [CrossRef]
16. Rembiesa, J.; Ruzgas, T.; Engblom, J.; Holefors, A. The impact of pollution on skin and proper efficacy testing for anti-pollution claims. *Cosmetics* **2018**, *5*, 4. [CrossRef]
17. Braga, P.C.; Ceci, C.; Marabini, L.; Nappi, G. The antioxidant activity of sulphurous thermal water protects against oxidative DNA damage: A comet assay investigation. *Drug Res.* **2013**, *63*, 198–202. [CrossRef] [PubMed]
18. Braga, P.C.; Sambataro, G.; Sasso, M.D.; Culici, M.; Alfieri, M.; Nappi, G. Antioxidant effect of sulphurous thermal water on human neutrophil bursts: Chemiluminescence evaluation. *Respiration* **2008**, *75*, 193–201. [CrossRef] [PubMed]
19. Prandelli, C.; Parola, C.; Buizza, L.; Delbarba, A.; Marziano, M.; Salvi, V.; Zacchi, V.; Memo, M.; Sozzani, S.; Calza, S.; et al. Sulphurous thermal water increases the release of the anti-inflammatory cytokine IL-10 and modulates antioxidant enzyme activity. *Int. J. Immunopathol. Pharmacol.* **2013**, *26*, 633–646. [CrossRef]
20. Tacheau, C.; Weisgerber, F.; Fagot, D.; Bastien, P.; Verdier, M.P.; Liboutet, M.; Sore, G.; Bernard, B.A. Vichy Thermal Spring Water (VTSW), a cosmetic ingredient of potential interest in the frame of skin ageing exposome: An in vitro study. *Int. J. Cosmet. Sci.* **2018**, *40*, 377–387. [CrossRef]
21. Rasmont, V.; Valois, A.; Gueniche, A.; Sore, G.; Kerob, D.; Nielsen, M.; Berardesca, E. Vichy volcanic mineralizing water has unique properties to strengthen the skin barrier and skin defenses against exposome aggressions. *J. Eur. Acad. Dermatol. Venereol.* **2022**, *36*, 5–15. [CrossRef]
22. Joly, F.; Gardille, C.; Barbieux, E.; Lefeuvre, L. eneficial effect of a thermal spring water on the skin barrier recovery after injury: Evidence for claudin-6 expression in human skin. *J. Cosmet. Dermatol. Sci. Appl.* **2012**, *2*, 273–276.
23. Mias, C.; Maret, A.; Gontier, E.; Carrasco, C.; Satge, C.; Bessou-Touya, S.; Coubetergues, H.; Bennett-Kennett, R.; Dauskardt, R.H.; Duplan, H. Protective properties of Avène Thermal Spring Water on biomechanical, ultrastructural and clinical parameters of human skin. *J. Eur. Acad. Dermatol. Venereol.* **2020**, *34*, 15–20. [CrossRef]
24. Zeichner, J.; Seite, S. From probiotic to prebiotic using thermal spring water. *J. Drugs Dermatol.* **2018**, *17*, 657–662.
25. Tamás, B.; Gabriella, K.; Kristóf; Anett, I.; Pál, K.J.; Bálint, T.; Péter, L.; Márton, P.; Katalin, N. The Effects of Lakitelek Thermal Water and Tap Water on Skin Microbiome, a Randomized Control Pilot Study. *Life* **2023**, *13*, 746. [CrossRef]
26. Figueiredo, A.C.; Rodrigues, M.; Mourelle, M.L.; Araujo, A.R.T.S. Thermal Spring Waters as an Active Ingredient in Cosmetic Formulations. *Cosmetics* **2023**, *10*, 27. [CrossRef]
27. Aires-Barros, L.; Marques, J.M.; Graça, R.C. Elemental and isotopic geochemistry in the hydrothermal area of Chaves, Vila Pouca de Aguiar (northern Portugal). *Environ. Geol.* **1995**, *25*, 232–238. [CrossRef]
28. Vaz, M.; Fernandes, P.O.; Ferreira, F.A.; Alves, M.J.; Costa, V.; Nunes, A. The importance-satisfaction matrix as a strategic tool for Termas de Chaves thermal spa priority improvements. *J. Tour. Sustain. Well-Being* **2023**, *11*, 52–65.
29. Clesceri, L.S. *Standard Methods for Examination of Water and Wastewater*; American Public Health Association: Washington, DA, USA, 1998; Volume 9.
30. *ISO 6222:1999*; Water Quality—Enumeration of Culturable Micro-Organisms—Colony Count by Inoculation in a Nutrient Agar Culture Medium. International Organization for Standardization: Geneva, Switzerland, 1999.
31. Gonçalves, B.; Falco, V.; Moutinho-Pereira, J.; Bacelar, E.; Peixoto, F.; Correia, C. Effects of elevated CO_2 on grapevine (*Vitis vinifera* L.): Volatile composition, phenolic content, and in vitro antioxidant activity of red wine. *J. Agric. Food Chem.* **2009**, *57*, 265–273. [CrossRef]
32. *ISO 10993-5*; Biological Evaluation of Medical Devices; Part 5: Tests for In Vitro Cytotoxicity. International Organization for Standardization: Geneva, Switzerland, 2009.
33. Carvalho, M.J.; Pedrosa, S.S.; Mendes, A.; Azevedo-Silva, J.; Fernandes, J.; Pintado, M.; Oliveira, A.L.S.; Madureira, A.R. Anti-Aging Potential of a Novel Ingredient Derived from Sugarcane Straw Extract (SSE). *Int. J. Mol. Sci.* **2023**, *25*, 21. [CrossRef] [PubMed]
34. Carvalho, M.J.; Pinto-Ribeiro, I.; Castro, C.; Pedrosa, S.S.; Oliveira, A.L.; Pintado, M.; Madureira, A.R. Impact of a novel sugarcane straw extract-based ingredient on skin microbiota via a new preclinical in vitro model. *Microbe* **2023**, *1*, 100017. [CrossRef]
35. Ryu, Y.S.; Kang, K.A.; Piao, M.J.; Ahn, M.J.; Yi, J.M.; Hyun, Y.-M.; Kim, S.H.; Ko, M.K.; Park, C.O.; Hyun, J.W. Particulate matter induces inflammatory cytokine production via activation of NFκB by TLR5-NOX4-ROS signaling in human skin keratinocyte and mouse skin. *Redox Biol.* **2019**, *21*, 101080. [CrossRef]
36. Byrd, A.L.; Belkaid, Y.; Segre, J.A. The human skin microbiome. *Nat. Rev. Microbiol.* **2018**, *16*, 143–155. [CrossRef]
37. Seite, S. Thermal waters as cosmeceuticals: La Roche-Posay thermal spring water example. *Clin. Cosmet. Investig. Dermatol.* **2013**, *6*, 23–28. [CrossRef]
38. Merial-Kieny, C.; Castex-Rizzi, N.; Selas, B.; Mery, S.; Guerrero, D. Avène Thermal Spring Water: An active component with specific properties. *J. Eur. Acad. Dermatol. Venereol.* **2010**, *25*, 2–5. [CrossRef]
39. Ferreira, M.O.; Costa, P.C.; Bahia, M.F. Effect of *São Pedro do Sul* thermal water on skin irritation. *Int. J. Cosmet. Sci.* **2010**, *32*, 205–210. [CrossRef] [PubMed]
40. Almeida, C.; Madeira, A.; Marto, J.; Graça, A.; Pinto, P.; Ribeiro, H. Monfortinho thermal water-based creams: Effects on skin hydration, psoriasis, and eczema in adults. *Cosmetics* **2019**, *6*, 56. [CrossRef]
41. Ali, S.M.; Yosipovitch, G. Skin pH: From basic science to basic skin care. *Acta Derm. Venereol.* **2013**, *93*, 261–267. [CrossRef]
42. Lukić, M.; Pantelić, I.; Savić, S.D. Towards optimal ph of the skin and topical formulations: From the current state of the art to tailored products. *Cosmetics* **2021**, *8*, 69. [CrossRef]

43. Nasermoaddeli, A.; Kagamimori, S. Balneotherapy in medicine: A review. *Environ. Health Prev. Med.* **2005**, *10*, 171–179. [CrossRef] [PubMed]
44. Casas, C.; Ribet, V.; Alvarez-Georges, S.; Sibaud, V.; Guerrero, D.; Schmitt, A.; Redoulès, D. Modulation of Interleukin-8 and staphylococcal flora by Avène hydrotherapy in patients suffering from chronic inflammatory dermatoses. *J. Eur. Acad. Dermatol. Venereol.* **2011**, *25*, 19–23. [CrossRef]
45. Castex-Rizzi, N.; Charveron, M.; Merial-Kieny, C. Inhibition of TNF-alpha induced-adhesion molecules by Avène Thermal Spring Water in human endothelial cells. *J. Eur. Acad. Dermatol. Venereol.* **2010**, *25*, 6–11. [CrossRef]
46. Nunes, F.; Rodrigues, M.; Ribeiro, M.P.; Ugazio, E.; Cavalli, R.; Abollino, O.; Coutinho, P.; Araujo, A.R.T.S. Incorporation of Cró thermal water in a dermocosmetic formulation: Cytotoxicity effects, characterization and stability studies and efficacy evaluation. *Int. J. Cosmet. Sci.* **2019**, *41*, 604–612. [CrossRef]
47. Oliveira, A.S.; Vaz, C.V.; Silva, A.; Correia, S.; Ferreira, R.; Breitenfeld, L.; Martinez-De-Oliveira, J.; Palmeira-De-Oliveira, R.; Pereira, C.; Cruz, M.T.; et al. In vitro evaluation of potential benefits of a silica-rich thermal water (Monfortinho Thermal Water) in hyperkeratotic skin conditions. *Int. J. Biometeorol.* **2020**, *64*, 1957–1968. [CrossRef] [PubMed]
48. Nicoletti, G.; Corbella, M.; Jaber, O.; Marone, P.; Scevola, D.; Faga, A. Non-pathogenic microflora of a spring water with regenerative properties. *Biomed. Rep.* **2015**, *3*, 758–762. [CrossRef] [PubMed]
49. Portugal-Cohen, M.; Oron, M.; Merrik, E.; Ben-Amitai, D.; Yogev, H.; Zvulunov, A. A Dead Sea Water-Enriched Body Cream Improves Skin Severity Scores in Children with Atopic Dermatitis. *J. Cosmet. Dermatol. Sci. Appl.* **2011**, *1*, 99–106. [CrossRef]
50. Joly, F.; Branka, J.-E.; Lefeuvre, L. Thermal Water from Uriage-les-Bains Exerts DNA Protection, Induction of Catalase Activity and Claudin-6 Expression on UV Irradiated Human Skin in Addition to Its Own Antioxidant Properties. *J. Cosmet. Dermatol. Sci. Appl.* **2014**, *04*, 99–106. [CrossRef]
51. Knott, A.; Drenckhan, A.; Reuschlein, K.; Lucius, R.; Döring, O.; Böttger, M.; Stäb, F.; Wenck, H.; Gallinat, S. Corrigendum to Decreased fibroblast contractile activity and reduced fibronectin expression are involved in skin photoaging. *J. Dermatol. Sci.* **2010**, *58*, 232. [CrossRef]
52. Sun, W.; He, J.; Zhang, Y.; He, R.; Zhang, X. Comprehensive functional evaluation of a novel collagen for the skin protection in human fibroblasts and keratinocytes. *Biosci. Biotechnol. Biochem.* **2023**, *87*, 724–735. [CrossRef]
53. Mormile, I.; Tuccillo, F.; Della Casa, F.; D'aiuto, V.; Montuori, N.; De Rosa, M.; Napolitano, F.; de Paulis, A.; Rossi, F.W. The Benefits of Water from Nitrodi's Spring: The In Vitro Studies Leading the Potential Clinical Applications. *Int. J. Mol. Sci.* **2023**, *24*, 13685. [CrossRef]
54. Grether-Beck, S.; Mühlberg, K.; Brenden, H.; Felsner, I.; Brynjolfsdottir, A.; Einarsson, S.; Krutmann, J. Bioactive molecules from the Blue Lagoon: In vitro and in vivo assessment of silica mud and microalgae extracts for their effects on skin barrier function and prevention of skin ageing. *Exp. Dermatol.* **2008**, *17*, 771–779. [CrossRef]
55. de Araújo, L.A.; Addor, F.; Campos, P.M.B.G.M. Use of silicon for skin and hair care: An approach of chemical forms available and efficacy. *An. Bras. de Dermatol.* **2016**, *91*, 331–335. [CrossRef]
56. Joly, F.; Charveron, M.; Aries, M.F.; Bidault, J.; Kahhak, L.; Beauvais, F.; Gall, Y. Effect of Avène Spring Water on the Activation of Rat Mast Cell by Substance P or Antigen. *Ski. Pharmacol. Appl. Ski.Physiol.* **1998**, *11*, 111–116. [CrossRef]
57. Berardesca, E.; Loden, M.; Serup, J.; Masson, P.; Rodrigues, L.M. The revised EEMCO guidance for the in vivo measurement of water in the skin. *Ski. Res. Technol.* **2018**, *24*, 351–358. [CrossRef] [PubMed]
58. Montero-Vilchez, T.; Segura-Fernández-Nogueras, M.-V.; Pérez-Rodríguez, I.; Soler-Gongora, M.; Martinez-Lopez, A.; Fernández-González, A.; Molina-Leyva, A.; Arias-Santiago, S. Skin Barrier Function in Psoriasis and Atopic Dermatitis: Transepidermal Water Loss and Temperature as Useful Tools to Assess Disease Severity. *J. Clin. Med.* **2021**, *10*, 359. [CrossRef] [PubMed]
59. Monteiro, R.C.; Nikam, V.N.; Dandakeri, S.; Bhat, R.M. Transepidermal Water Loss in Psoriasis: A Case-control Study. *Indian Dermatol. Online J.* **2019**, *10*, 267–271. [CrossRef] [PubMed]
60. Martin, R.; Henley, J.B.; Sarrazin, P.; Seité, S. Skin Microbiome in Patients With Psoriasis Before and After Balneotherapy at the Thermal Care Center of La Roche-Posay. *J. Drugs Dermatol.* **2015**, *14*, 1400–1405. [PubMed]
61. Haftek, M.; Abdayem, R.; Guyonnet-Debersac, P. Skin Minerals: Key Roles of Inorganic Elements in Skin Physiological Functions. *Int. J. Mol. Sci.* **2022**, *23*, 6267. [CrossRef]
62. Woolery-Lloyd, H.; Andriessen, A.; Day, D.; Gonzalez, N.; Green, L.; Grice, E.; Henry, M. Review of the microbiome in skin aging and the effect of a topical prebiotic containing thermal spring water. *J. Cosmet. Dermatol.* **2023**, *22*, 96–102. [CrossRef]
63. Manara, S.; Beghini, F.; Masetti, G.; Armanini, F.; Geat, D.; Galligioni, G.; Segata, N.; Farina, S.; Cristofolini, M. Thermal Therapy Modulation of the Psoriasis-Associated Skin and Gut Microbiome. *Dermatol. Ther.* **2023**, *13*, 2769–2783. [CrossRef]
64. Instituto Nacional de Investigação Agrária e Veterinária (INIAV)Estudo do Microbismo Natural das Águas Minerais Naturais. 2018. Available online: https://hidrogenoma.dgeg.gov.pt/agua-mineral-natural/caldas-de-chaves (accessed on 27 May 2024).

applied sciences

MDPI

Article

A Comprehensive Assessment of Sample Preparation Methods for the Determination of UV Filters in Water by Gas Chromatography–Mass Spectrometry: Greenness, Blueness, and Whiteness Quantification Using the AGREEprep, BAGI, and RGB 12 Tools

Grażyna Wejnerowska and Izabela Narloch *

Department of Food Analysis and Environmental Protection, Faculty of Chemical Technology and Engineering, Bydgoszcz University of Science and Technology, 85-326 Bydgoszcz, Poland; grazyna.wejnerowska@pbs.edu.pl
* Correspondence: izabela.narloch@pbs.edu.pl

Abstract: Sample preparation is a key step in the analytical procedure. This step is a time- and labor-consuming process, and often it is also expensive, with costs being influenced by the consumption of materials and reagents. Additionally, the toxicity of the reagents, waste generation, and energy consumption affect the environment and the safety of the analyst. New trends in sample preparation are focused on the development of miniaturized methods that are consistent with the principles of green sample preparation and contribute to environmental sustainability. The results of a comprehensive assessment of ten methods of preparing water samples for the determination of UV filters using gas chromatography are presented. Three assessment tools were used for this purpose: AGREEprep (the analytical greenness metric for sample preparation), BAGI (the blue applicability grade index), and the RGB 12 algorithm (red–green–blue model). All the differences and similarities between the three aforementioned metrics are discussed in this manuscript. The results of the evaluation of the most frequently used microextraction methods show their ecological friendliness, effectiveness, and practicality. The results of this assessment will allow researchers to identify the strengths and weaknesses of the given methods and select those that meet their requirements.

Keywords: assessment tools; gas chromatography; greenness; microextraction methods; UV filters

1. Introduction

UV filters are a group of chemicals commonly used in a wide range of cosmetic products to protect the skin from the harmful effects of UV radiation [1,2]. Organic UV filters have a highly lipophilic character, and most of them are classified as water-resistant and therefore tend to accumulate in the fatty tissues of living organisms [3,4]. They are considered emerging contaminants since they easily enter the natural environment, where they accumulate, causing harmful effects on flora and fauna despite being present at the ng/L level. Therefore, developing sensitive and selective analytical methods for their environmental monitoring is of high interest [5]. As exemplified by the Web of Science database, the results for the combination keywords "UV filters" and "environmental water" showed a growing trend in the amount of research (from 2002 to 2023) on the contamination of the waters by UV filters (Figure 1).

Based on the results obtained from the database, an increasing number of applied microextraction techniques compared to conventional techniques can be observed. The most applied classic extraction methods include solid-phase extraction (SPE), liquid–liquid extraction (LLE), fabric phase sorptive extraction (FPSE), Quick, Easy, Cheap, Effective, Rugged, and Safe Extraction (QuEChERS), and magnetic nanoparticles dispersive solid-phase extraction (MNPs-based dSPE). Microextraction techniques include solid-phase microextraction (SPME), stir bar sorptive extraction (SBSE), dispersive solid-phase extraction (dSPE), microextraction by packed sorbent (MEPS), bar adsorptive microextraction (BAμE), stir bar sorptive dispersive microextraction (SBSDME), single-drop microextraction (SDME), in situ suspended aggregate microextraction (iSAME), hollow-fibre liquid-phase microextraction (HFLPME), dispersive liquid–liquid microextraction (DLLME), ultrasounds-assisted dispersive liquid–liquid microextraction (USA-DLLME), vortex-assisted dispersive liquid–liquid microextraction (VA-DLLME), and ultrasounds-assisted emulsification microextraction (USAE-ME). Based on the literature review, it is concluded that the most popular microextraction techniques for determining UV filters are the following: SPME (~24%); DLLME, with various variants (~24%); SBSE (~16%); MSPE (~5%); and others (~16%). For instrumental techniques, the most common choices for the detection and quantification of the compounds studied in water samples remain gas chromatography and liquid chromatography, coupled with mass spectrometry [5–8].

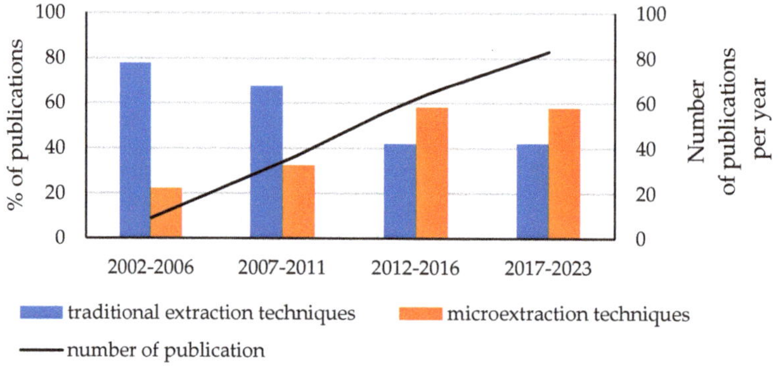

Figure 1. Evolution of the number of publications (%) concerning the determination of UV filters in environmental water samples (2002–2023) by using traditional extraction techniques and microextraction techniques (188 articles on the determination of UV filters in water samples).

Selecting the most appropriate analytical method for determining UV filters in water samples, among many developed methods, is not an easy task. When choosing a method, validation criteria should be taken into account, i.e., accuracy, precision, sensitivity, and selectivity, as well as economic and practical factors, i.e., costs, time, and ease of use. Moreover, special attention is currently being paid to the ecological aspect of the analytical method. According to the principles of Green Analytical Chemistry (GAC), methods should be used that do not pose a threat to human health and the environment.

GAC is an aspect of green analytical chemistry that was introduced in the late 1990s. It is a concept that is based on twelve principles related to the environment, health, and safety [9]. GAC takes into account, among other things, the use of safe solvents/reagents, the generation of toxic waste, and the safety of the analysts. In 2022, López-Lorente et al. [10] proposed ten principles of Green Sample Preparation (GSP) that aim to develop greener analytical procedures. In GSP, the GAC principles have been extended to include the use of solvents/reagents from renewable sources and reusable and/or recyclable materials. Additionally, GSP takes into account sample throughput, miniaturization, and the automation of the method. However, in 2021, Nowak et al. [11] introduced a new concept of sustainable development in analytical chemistry, the so-called White Analytical Chemistry (WAC), which is an extension of GAC. The authors of WAC proposed the WAC principles as an alternative to the 12 GAC principles but including not only green aspects (such as the toxicity of reagents, the number and amount of reagents and waste, energy, and other media, as well as other direct impacts). WAC also takes into account criteria such as analytical efficiency (scope of application, limits of quantification and detection, precision, and accuracy) and practical/economic criteria (cost-efficiency, time-efficiency, requirements, and operational simplicity). The compliance of analytical methods with the GAC, GSP, and WAC principles is a basic requirement in the development of sustainable analytical methods.

Over the last few years, various metric tools have been introduced to assess the environmental performance of analytical methods, including the National Environmental Method Index (NEMI) [12], the Analytical Eco-Scale [13], the Green Analytical Procedure Index (GAPI) [14], the Analytical Greenness Calculator (AGREE) [15], the RGB 12 algorithm [11], the Analytical Method Greenness Score (AMGS) [16], the Blue Applicability Grade Index (BAGI) [17], the Complementary Green Analytical Procedure Index (Complex-GAPI) [18], and the Analytical Greenness Metric for Sample Preparation (AGREEprep) [19]. All of these tools graphically and/or numerically reflect the compliance of a given analytical method with the GAC principles.

The main aim of this work was to assess the ecological and practical aspects of the water sample preparation step for the determination of UV filters by GC-MS using the AGREEprep, BAGI, and RGB 12 tools. These are the newest tools that are most often chosen for the evaluation of analytical procedures due to their versatility, usefulness, and simplicity of use.

Using these three metrics simultaneously will provide comprehensive information about the strengths and weaknesses of the analytical procedures used. At the same time, the presented correlations between the used metrics can be a guide for the analyst when deciding on the selection of a metric tool.

The assessment of analytical methods is necessary to understand their impact on the environment and can be helpful for analytical chemists when choosing a method for determining cosmetic ingredients and other compounds in water samples.

2. Materials and Methods

2.1. AGREEprep

AGREEprep is a new analytical "greenness" metric that was published by Wojnowski et al. [19] in 2022. The free version of the software can be obtained from https://mostwiedzy.pl/AGREE (accessed on 3 May 2024) [20]. AGREEprep is based on ten steps of assessment that correspond to the ten principles of GSP. In order to assess the "greenness" of an analytical method, AGREEprep is based on ten individual steps which are presented in Table 1. Each criterion is scored from 0 to 1, with the extremes representing the worst and best performance, respectively. Moreover, each criterion has a default weight taken into account in the overall score, but the assessors can change this value at their discretion if there are valid reasons to do so [20]. However, in our work, we did not change the value of the criteria because we considered the default weights assigned to the criteria to be correct.

The result of the AGREEprep assessment is a colorful round pictogram that maps the degree of compliance of evaluated criteria within the rules of GAC. The color of the circle inside the pictogram and the overall score within it indicate the overall environmental performance of the sample preparation in a given analytical method. The overall score can range from 0 to 1, with a score of 0 being the worst result and 1 being the best result, taking into account the scores from all criteria or the lack of a sample-preparation step. On the outer part of the circle, there are ten parts, corresponding to the ten criteria. Each part may have a different length, depending on the weight assigned to a given criterion. However, the color of a given part indicates its score—a highest criterion score is indicated in green, and a lowest criterion score is indicated in red. A result between 0 and 1 is represented by a color gradient between red and green, e.g., yellow and orange in different shades, ac-cording to the value assigned by the AGREEprep calculator.

Table 1. Description of the criteria and graphical presentation of results for AGREEprep, WAC, and BAGI metrics.

AGREEprep	WAC	BAGI
1. Favor in situ sample preparation 2. Use safer solvents and reagents 3. Target sustainable, reusable, and renewable materials 4. Minimize waste 5. Minimize sample, chemical, and material amounts 6. Maximize sample throughput 7. Integrate steps and promote automation 8. Minimize energy consumption 9. Choose the greenest possible post-sample preparation configuration for analysis 10. Ensure safe procedures for the operator	RED R1: Scope of application R2: LOD and LOQ R3: Precision R4: Accuracy GREEN G1: Toxicity of reagents G2: Amount of reagents and waste G3: Energy and other media G4: Direct impacts BLUE B1: Cost-efficiency B2: Time-efficiency B3: Requirements B4: Operational simplicity	1. The type of analysis 2. The number of analytes that are simultaneously determined 3. The analytical technique and required analytical instrumentation 4. The number of samples that can be simultaneously treated 5. Sample preparation 6. The number of samples that can be analyzed per hour 7. The type of reagents and materials used in the analytical method 8. The requirement for preconcentration 9. The automation degree 10. The amount of sample
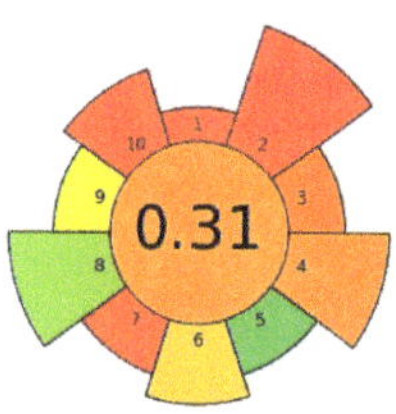	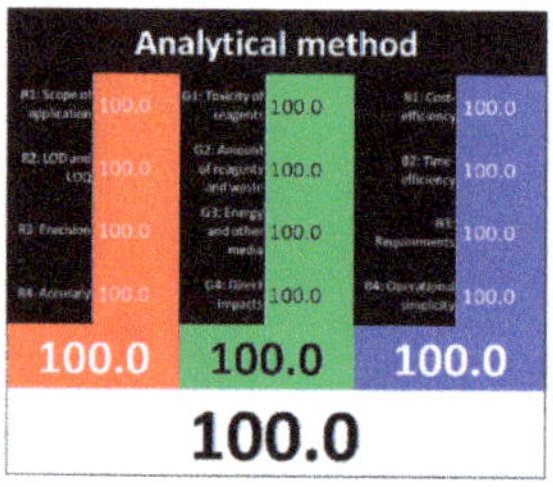	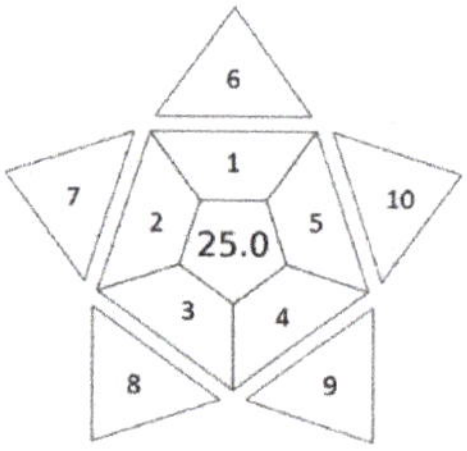

2.2. WAC

White analytical chemistry (WAC) is a concept of sustainable development in analytical chemistry, which is an extension of green analytical chemistry. WAC was designed and developed by Nowak et al. [11] in 2021, and it is a concept that encourages the harmony and integration of analytical, ecological, and practical characteristics, while aiming for the sustainability of analytical methods. In the WAC concept, the RGB (red, green, blue) model [21] is used to evaluate the analytical method. Just as the color white is created by mixing red, green, and blue light, the analytical method becomes white, and thus complete, when it achieves each primary color.

To evaluate the methods using the RGB 12 algorithm, which is the second version of the RGB model adapted to the 12 criteria of WAC, the available Excel template spreadsheet is used (access in Supplementary data in Nowak et al.'s work [11]), where specially prepared tables of red, green, and blue colors can be found. The template was designed to be able to evaluate and compare 10 methods simultaneously. The tables should be completed by assigning each criterion a point value ranging from 0 to 100. A value of 0 means the worst result, and 100 means that the method is well suited to the planned application. It is also possible to award more than 100 points for outstanding criteria in the evaluation of an analytical method. After completing the form, the assessment results are automatically calculated and presented in tabular form. The compliance of the method with a given WAC criterion is presented both numerically and visually by saturating a given value with color (a criterion value of 0 corresponds to black; 100 or more points correspond to full-color saturation). The value of arithmetic means for the red, green, and blue criteria are expressed individually as R (%), G (%), and B (%), while the overall result (whiteness—%) is given in the table and figure (Table 1).

2.3. BAGI

The blue applicability grade index (BAGI) is a new analytical "blueness" metric tool for evaluating the practicality of an analytical method, which was published by Manousi et al. [17] in 2023. The free version of the software can be obtained from https://mostwiedzy.pl/pl/justyna-plotka-wasylka,647762-1/BAGI (accessed on 3 February 2024). The blue color in the BAGI metric is inspired by the RGB model, and it may be considered complementary to the existing green metrics tools. In order to assess the applicability of an analytical method, BAGI takes into account the criteria shown in Table 1.

The overall result of assessing the method using BAGI is an asteroid pictogram with a number in the center. The hue of the scale of the pictogram reflects the compliance of the method with the designated criteria. There are four colors in the BAGI: dark blue for high compliance, blue for medium compliance, light blue for low compliance, and white for no compliance. The number in the center of the pictogram indicates the overall score for the analytical method, which is a number ranging from 25 to 100. A point value of 100 is assigned to a method with excellent performance, and a value of 25 indicates the worst performance of the method in terms of applicability. A method whose BAGI score is at least 60 points is considered practical. In the pictogram, criteria 1–5, located in its inner part, correspond to the stage of analytical determination or sample preparation stage. However, criteria 6–10, placed in the outer part, correspond to both mentioned stages. The result field takes on a shade that is the average shade of all criteria taken into account in BAGI.

3. Results and Discussion

3.1. Greenness, Blueness, and Whiteness Evaluation

The paper presents an assessment of the environmental impact and analytical suitability of ten methods of preparing water samples for the determination of UV filters described in the literature. One of the methods, solid phase extraction (SPE), is a classic extraction method, while the other nine are commonly used microextraction techniques. Table 2 presents assessed analytical methods and their literature sources. A brief description of the assessed sample preparation methods for analysis is presented in Supporting Information (Table S1).

Table 2. Results from the evaluation of methods for preparing water samples for the analysis of UV filters obtained using the AGREEprep, BAGI, and WAC metrics.

Method	AGREEprep	WAC [%] *	BAGI
SPE—solid phase extraction [22]	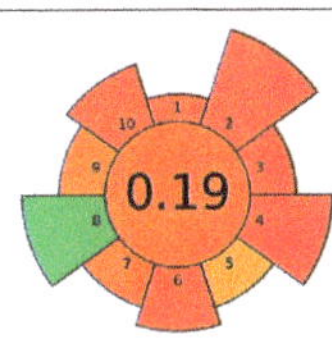	R 91.3 G 75.0 B 66.3 W 77.5	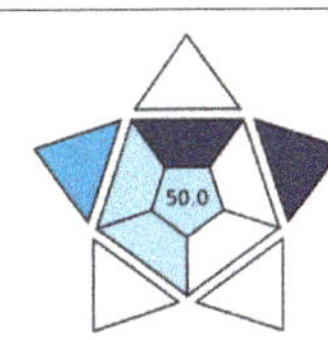
SBSE—stir bar sorptive extraction [23]	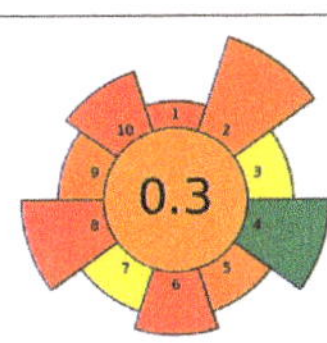	R 83.8 G 81.7 B 51.5 W 72.3	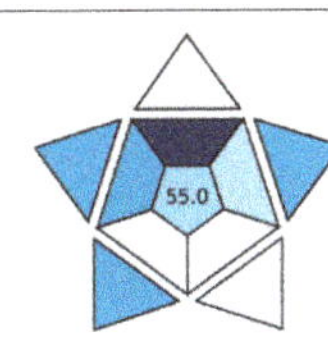
USAEME—ultrasound-assisted emulsification microextraction [24]	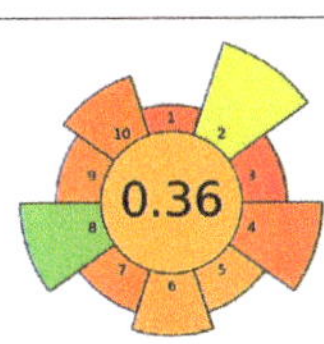	R 92.5 G 86.7 B 92.9 W 90.7	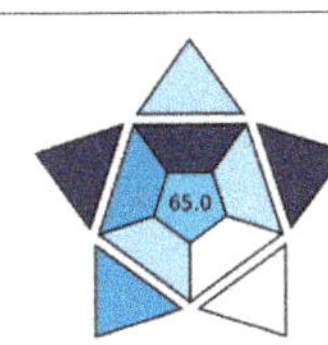

Table 2. *Cont.*

Method	AGREEprep	WAC [%] *	BAGI
DLLME—dispersive liquid-liquid microextraction [25]	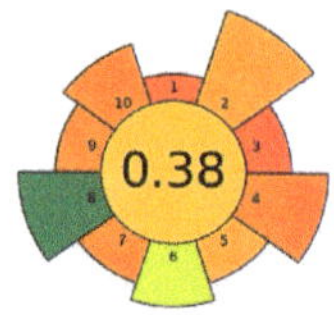	R 90.0 G 90.8 B 100.4 W 93.8	
SPME (derivatization)—solid-phase microextraction (with derivatization) [26]	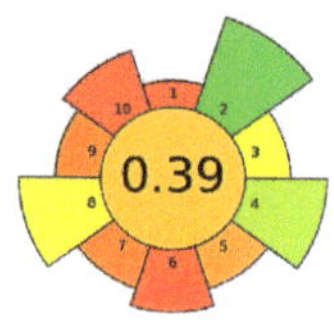	R 95.0 G 96.3 B 67.1 W 86.1	
MEPS (automated)—microextraction by packed sorbent [27]	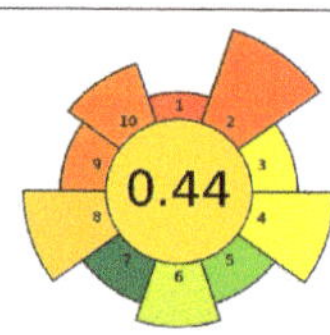	R 87.5 G 89.6 B 87.9 W 88.3	
MEPS—microextraction by packed sorbent [28]	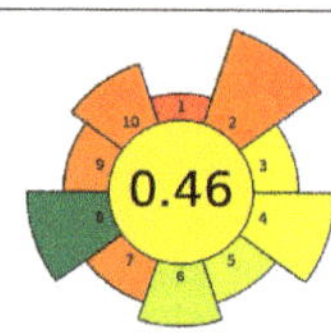	R 91.3 G 98.8 B 97.1 W 95.7	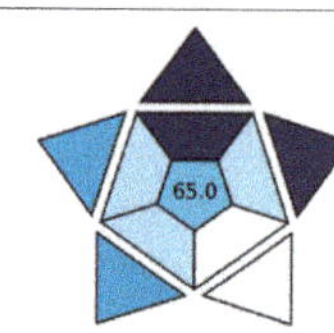
DmSPE—dispersive micro solid-phase extraction [29]	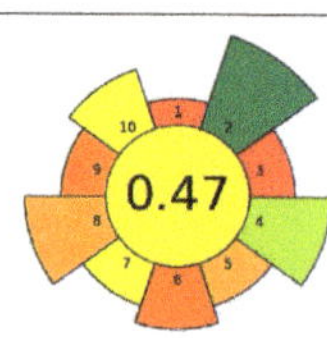	R 97.5 G 99.6 B 80.2 W 92.4	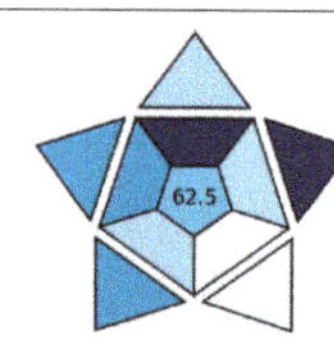
SDME—single-drop microextraction (named by authors of article liquid-phase microextraction [30])	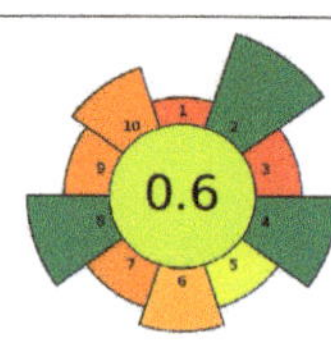	R 95.0 G 101.3 B 103.8 W 100.0	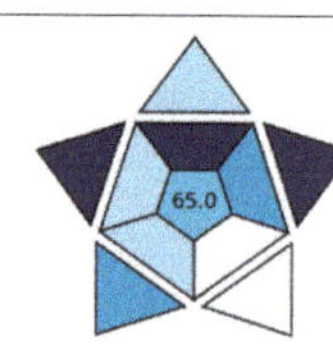
SPME—solid-phase microextraction [31]	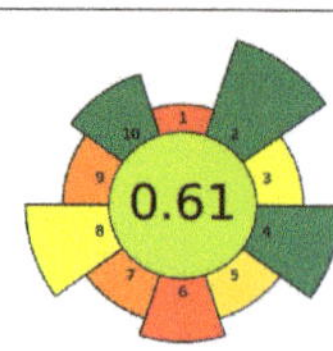	R 93.8 G 106.7 B 72.1 W 90.8	

* R—red principles (analytical performance); G—green principles (green chemistry); B—blue principles (practical side); W—whiteness.

All selected sample preparation procedures use gas chromatography and mass detection (GC-MS) for analysis. This made it possible to evaluate only the sample preparation step without taking into account the time, costs, and energy needed to perform the chromatographic analysis. In the case of methods using thermal desorption of analytes from the sorbent in the chromatograph dispenser, i.e., SBSE and SPME, the desorption step was included in the total assessment (time and energy consumption). It was also assumed that one sample was analyzed, and the time and costs incurred for its optimization were not taken into account when assessing the method. Using the example of determining UV filters in water samples, the environmental friendliness and functionality of commonly used microextraction methods were assessed using three tools, i.e., AGREEprep, BAGI, and RGB 12. AGREEprep assesses in detail the environmental impact of the use of a given analytical procedure, BAGI assesses its usefulness, while RGB 12 covers its comprehensive assessment ecological performance, analytical performance, and its economic and practical aspects.

3.1.1. AGREEprep Assessment

The use of AGREEprep allows the critical evaluation of each step of sample preparation from the GAC point of view. The results of the assessment performed using AGREEprep are presented in Table 2. As expected, the lowest score (0.19) was obtained by the SPE method, in which only one of the principles, energy use, was green (criterion 8). Classic extraction methods, including SPE, are characterized by high consumption of toxic solvents (criterion 2 uses safer solvents, and criterion 10 is safe for the operator), generate large amounts of waste (criterion 4), and are time-consuming (criterion 6). The SPE method, among the classical methods, is the most frequently used method for analyzing UV filters, which is the reason for presenting it in this work to compare it with microextraction methods.

However, the lowest AGREEprep score among the presented microextraction methods was given to the SBSE method, whose score was 0.3. Such a low assessment of the greenness of this method is due to the long time (criterion 6) of extraction (180 min) and desorption (15 min), as well as the energy (criterion 8) consumed during sample mixing, especially during the desorption step.

The USAEME method received a low AGREEprep score of 0.36. Only 100 μL of chloroform was used for extraction, which has a high score effect on the assessment of criterion 2. In this procedure, a large amount of sample was used (10 mL; criterion 5), to which 2 g of NaCl was added, together with the solvent. Due to the solvent contamination of the aqueous solution, all of the solution was treated as waste, which significantly lowered the score for criterion 4.

Sample preparation using the DLLME method takes very little time (~5 min), which results in a favorable result for criterion 6 (sample throughput). In this method, the sample is subjected to a short centrifugation (3 min), thanks to which the energy consumption is low, which is why criterion 8 has a green rating. However, the use of two hazardous solvents (1 mL acetone and 60 μL chlorobenzene) negatively affects criteria 2 and 10 (use safer solvents and safe for the operator), and it also results in a low AGREEprep score of 0.38.

The UV filters contain phenolic hydroxyl groups, which cause the low sensitivity of the GC analysis. The use of derivatization increases sensitivity and improves separation and shape peaks. Therefore, derivatization is often used in their analysis. Derivatization (on-fiber silylation) was used in the next assessed method—SPME. However, its use has a negative impact on the assessment of the ecological effectiveness of the method. The SPME with the derivatization method received a score of 0.39 in the AGREEprep evaluation. The reagents used for sample acidification and derivatization (MSTFA), and the time needed to perform derivatization, extraction, and desorption (~45 min) reduce the scores for criteria 6 and 10 (sample throughput and safety for the operator). Additionally, the energy consumed for mixing, sample heating, and thermal desorption reduces the score for criterion 8. For comparison, the SPME procedure (without derivatization) obtained the highest result of 0.61 among the assessed methods. A much smaller impact on the reduction of the greenness rating due to derivatization was demonstrated for other methods, i.e., DLLME (ultrasound-assisted) [32] and SBSE [33]. The AGREEprep rating for the DLLME method decreased from 0.38 to 0.3, and for SBSE from 0.3 to 0.28. In these methods, unlike SPME, derivatization is carried out simultaneously with extraction, which does not result in increased time and energy consumption.

The example of the MEPS method shows the differences in the AGREEprep assessment for the fully automated and manual MEPS method, which received scores of 0.44 and 0.46, respectively. Both methods have one green criterion: the manual MEPS scores green on criterion 8 (energy consumption), while the fully automated method scores green on criterion 7 (integration, automation). However, criterion 8 has a higher weight in the AGREEprep metric, which causes a difference in the evaluation of both methods. The rating of both MEPS techniques is not the highest, influenced by the fact that solvents are used for extraction (low criterion 2), and it additionally reduces the criteria related to waste generation and safety for the operator (criteria 4 and 10).

Dispersive micro- solid-phase extraction (DmSPE) is a solvent-free method, with a positive impact on criterion 2, which received the green status. However, the total score for this procedure is not high, and amounted to 0.47. The derivatization step of analytes reduces the overall evaluation of the method. After extraction, the analytes, together with the sorption bed, are placed in the injection port, where they are derivatized and then thermally desorbed. The use of a derivatization reagent (BSTFA) extends the sample preparation time (criterion 6—sample throughput) and reduces criteria 2 and 10 (use safer reagents and operator safety).

Only two microextraction methods achieved the green status, with scores of 0.6 and 0.61—SDME and DI-SPME, respectively. The SDME method obtained such a high rating compared to the previously discussed methods thanks to the use of a very small amount of solvent (3 µL of toluene) for extraction, which is injected into the injection port after extraction. This affects two green criteria—criterion 2 (use safer solvents) and criterion 4 (minimize waste). The favorable final assessment was also influenced by the use of only 2 mL of sample for analysis (criterion 5) and the consumption of a small amount of energy (criterion 8) due to magnetic stirring for 20 min. Compared to SDME, criterion 8 in the SPME method has a lower score due to the higher energy consumption during the thermal desorption step and the long operating time of the magnetic stirrer during the sorption step (45 min). However, because it is a solvent-free, waste-free method and no derivatization of analytes was used, three criteria obtained a green rating, with a result of 1 (criteria 2, 4, and 10—use safer reagents, minimize waste and operator safety), which resulted in the highest score (0.61) for the SPME method in terms of greenness.

3.1.2. WAC Assessment

The RGB 12 metric, compared to the AGREEprep and BAGI metrics, allows for a deeper analysis of the method in terms of its analytical performance and practical benefits. This metric allowed the authors to assign scores to individual categories according to their discretion (experience and needs). All data of the assessed procedures were compiled in the available Excel spreadsheet, and, taking into account the importance of each criterion, points were assigned to all parameters.

The scores obtained for each principle are presented in Table 2 and Figure 2, and the detailed elements of this assessment are presented in an Excel spreadsheet (Supporting Information).

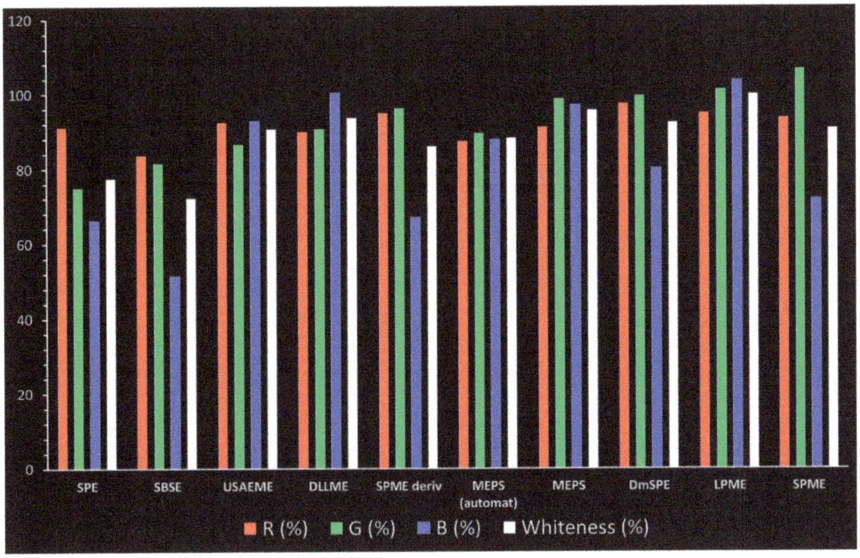

Figure 2. Comparison of the main assessment outcomes from the RGB 12 tool. Values above 100 indicate additional capabilities exceeding current requirements.

RED Principles Rating (Analytical Performance)

The red category evaluates the method in terms of its applicability to its intended purpose. Precision, accuracy, limit of detection, and scope of application are assessed in this category. When determining the UV filters in the environmental samples, this category is of great importance due to the presence of these compounds at very low concentration levels and the wide range of UV filters used.

High redness results of 95% were obtained by the SPME (with derivatization) and SDME methods, while the DmSPE method was rated highest in this category, with a score of 97.5%. Of the assessed parameter components (R2), the lowest LOD (0.5–10 ng/L) was achieved by the SPME (on-fiber silylation) and DmSPE methods, earning them 100 points. These methods also showed high precision and accuracy, which contributed to the high final results in this category.

However, the LODs of the remaining methods were at a similar concentration level. The last points in this category were awarded to the SPE (70%) and SPME (75%) methods, for which the LOD was ~1–8 μg/L. However, it should be taken into account that the LOD was determined in laboratories using MS detectors operating at various parameters.

Green Principles Rating (green chemistry). The authors of this work entered the data of all assessed procedures into the Excel spreadsheet for the following categories: G1: toxicity of reagents, G2: amount of reagents and waste, G3: consumption of energy, and G4: direct impacts (safety). Based on these data, they assigned points from 0 to 120 to individual categories. It was decided to award 120 points to categories that showed 0 (reagents, waste, energy).

The greenest methods according to WAC are SPME, with a score of 106.7%, and SDME, with a score of 101.3%. The SPME method was awarded 120 points for the G1, G2, and G4 criteria. This is because only in this procedure no solvents and reagents were used. However, the SDME procedure requires the use of only 2 μL of solvent, which also contributes to the high rating of this method. Neither method generates waste and both are safe for the operator. The greenness rating of the remaining microextraction methods ranged from 81.7 to 99.6. These differences mainly resulted from the amount of energy used. As expected, SPE received the lowest greenness score (75), which is mainly due to the large amount of reagents used (~40 mL) and waste generated (30 mL). These parameters mean that for the G2 principle, SPE received only 20 points.

BLUE Principles Rating (Practical Side)

The assessment of practical and economic aspects includes the following categories: B1: cost-efficiency, B2: time-efficiency, B3: requirements, and B4: operational simplicity. In the blue principles, it was decided to distinguish procedures by awarding them 120 points for zero financial contribution to equipment and reagents in category B1, for methods whose sample preparation time is lower than 5 min in category B2, and for methods using less than 1 mL of sample in category B3.

Two of the assessed procedures, SDME and DLLME, received over 100 points: 103.8 and 100.4, respectively. The main advantages of these methods are their low costs and speed of implementation (B1 and B2). The MEPS (manually) and USAEME methods also have high scores, of over 90 points. However, the remaining methods (SBSE, SPME, SPE), due to the costs of purchasing accessories, automatic attachments (e.g., thermal desorption for SBSE), and derivatization reagents, received a low rating for category B1. Additionally, the total rating of these methods was lowered by the B3 category, which is influenced by the requirements for "advanced instruments and greater operator skills and experience".

Whiteness

Whiteness is a summary assessment of the three components (red, green, and blue) and shows the overall usefulness of the procedure. This assessment tool enables the analyst to select a procedure that will meet the given expectations. The assessment results, together with knowledge of the matrix composition, amounts, properties, and expected concentrations of analytes, laboratory equipment, economic opportunities, and analyst skills will help to select the most appropriate procedure from among the highest-rated procedures.

The highest whiteness rating (Table 2, Figure 2) of 100% was awarded to SDME, with only the red principle rated at less than 100 points. Assuming that the authors of this paper consider a result of more than 90% to indicate the whiteness of the method, six out of nine microextraction methods achieved satisfactory results. However, the lowest whiteness ratings were received by SBSE (72.3%) and SPE (77.5%), for which the blue principles had the greatest impact on the assessment.

3.1.3. BAGI Assessment

BAGI evaluates ten main attributes of an analytical procedure in terms of practicality. The results of the assessment performed using BAGI are presented in Table 2. Six of the assessed methods obtained a high score, higher than 60, indicating their practicality. According to this assessment, the highest score was 70 points, obtained by the DLLME and MEPS (automatic) methods, then MEPS (manually), USAEME, and SDME—65 points—and DmSPE—62.5 points.

The remaining methods, i.e., SPE, SPME, and SBSE, received scores in the range of 50–60 points. This assessment is influenced, among other things, by parameter 7 (reagents and materials), which gives a low score for the need to purchase "commercially available reagents and materials" such as SPE cartridges, SPME fibers, SBSE twisters, and derivatization reagents. Additionally, the SBSE and MEPS (fully automated) methods received 0 points for parameter 3 (analytical technique) for "instrumentation that is not commonly available in most labs".

3.2. *AGREEprep vs. GREEN Principle of WAC Assessment*

It was verified whether the greenness assessment of the methods performed using the AGREEprep tool was consistent with the green principles of the WAC assessment. The methods assessed are presented in Table 2 in ascending order of score (from the AGREEprep assessment), and this order was in most cases confirmed after the assessment of the green principles of WAC. The only discrepancy was observed in the greenness assessment for the MEPS (fully automated) method, which obtained a higher result in the AGREEprep assessment. This higher score was related to criterion 7, where the automated method receives additional points in the AGREEprep assessment. However, in the RGB 12 algorithm, the automation of methods is assessed in the principle blue: 4 (operational simplicity). Despite this, a high degree of convergence was observed in the greenness assessments of both metrics used to evaluate the ten methods for preparing water samples for the analysis of UV filters using the GC-MS technique. This convergence is shown in Figure 3, which shows that the correlation of both assessments is high and amounts to 0.877.

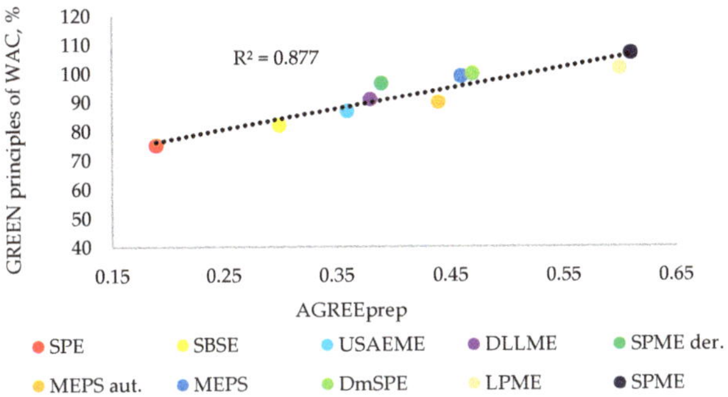

Figure 3. Linear correlation between the AGREEprep and GREEN principles of WAC.

3.3. BAGI vs. BLUE Principle of WAC Assessment

The BAGI tool is used to assess the suitability of an analytical method. However, the RGB 12 algorithm includes the blue principle (practical side), which is one of the components of the total whiteness assessment of the method. The correlation between the results obtained using the BAGI and the blue principle tools is shown in Figure 4. Despite a moderate correlation (R2 ~ 0.7), there was some agreement between both assessments. The four highest-rated methods for suitability by both tools are DLLME, SDME, MEPS, and USAEME. These methods obtained results for BAGI > 60 points, and for the blue principles > 90%. However, the SPME, SBSE, and SPE methods were rated the lowest by both tools. As can be seen in Figure 4, a greater discrepancy in both assessments can be seen for two methods: SPE and MEPS (fully automated). In the case of the SPE method, the BAGI rating is low (50), which is influenced by principle 8 (preconcentration), which lowers the rating for methods that use additional concentration steps (solvent evaporation). However, in the case of the MEPS (fully automated) method, the blue rating in relation to BAGI is lowered by additional parameters assessed in the blue metric tool, i.e., B1 (total cost) and B4 (portability).

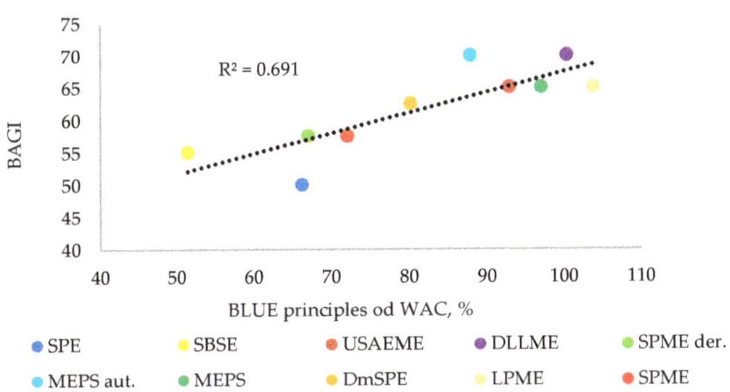

Figure 4. Linear correlation between BAGI and blue principles of WAC.

3.4. Summary of Evaluation of Sample Preparation Methods

The data presented in the Introduction indicate that the most popular microextraction techniques for determining UV filters are SPME (~24%) and DLLME, with various variants (~24%). In the case of the SPME method, this result is very consistent with the AGREEprep evaluation, in which this technique obtained the highest result (0.61). It is a solvent-free and waste-free method, which makes it unrivalled in the "green" category. However, it can be noticed that the best evaluation results in all categories (green, blue, and white) were obtained by the SDME method. However, these highest ratings for the SDME method do not translate into the popularity of using this method in practice (~3%). This is probably mainly related to maintaining a stable solvent microdrop. However, another popular method used by analysts—DLLME—is easy to perform and cheap. Its frequent use coincides with the highest BAGI rating (70 points), indicating its practicality.

These observations confirm the fact that analysts, using their experience, accurately select the most beneficial techniques. These techniques are simple, cheap, fast, and solvent-free or use minimal amounts of solvents. They are also characterized by reliability, repeatability, and sensitivity. Additionally, what is important is that their use does not require the purchase of additional laboratory equipment. However, the evaluation tools used confirm and prove the selection of the most advantageous analytical technique, and they are used to evaluate newly developed procedures.

4. Conclusions

The results of the assessment of sample preparation procedures presented in this work, based on the example of determining UV filters in water samples, demonstrated the usefulness and effectiveness of all metric tools used. The WAC tool was found to evaluate the methods most comprehensively, as expected. The advantage of this tool is that it allows the analyst to independently assign points for individual principles according to individual problems and needs, although this task is time-consuming and difficult to perform. However, the AGREEprep and BAGI tools evaluate the method within a narrower scope (greenness and practicality), and their implementation is relatively simple and quick. As a result of the evaluations, AGREEprep and BAGI were highly consistent with the WAC tool. There is no doubt that the procedure assessment tools used in this work are useful and help the analyst decide which method to choose for use in the laboratory.

As shown by the evaluation carried out using three complementary tools, the SDME, SPME, and DLLME methods were rated the highest. Two methods, i.e., SPME and SDME, obtained the highest greenness results which was confirmed by the green principles of WAC (>100%). However, the best method in the practicality category is DLLME, which received 70 points in the BAGI assessment, and it was also confirmed by the blue principles of WAC (>100%).

The presented assessment results show that the use of expensive materials and devices and conducting additional steps for the procedure, e.g., derivatization, sonication, etc., negatively affect all assessment. The preferred and still-developing methods should be (if possible) simple, cheap, fast, and preferably solvent-free. However, when choosing a method, it is necessary to remember to maintain a balance between greenness, functionality, and usability.

Supplementary Materials: The following supporting information can be downloaded at: https://www.mdpi.com/article/10.3390/app14177690/s1, Table S1: A brief description of the analytical methods assessed; Excel spreadsheet: WAC assessment.

Author Contributions: Conceptualization, G.W. and I.N.; formal analysis, G.W.; methodology, G.W. and I.N.; writing—original draft preparation, G.W. and I.N.; writing—review and editing, G.W. and I.N. All authors have read and agreed to the published version of the manuscript.

Funding: This research received no external funding.

Institutional Review Board Statement: Not applicable.

Informed Consent Statement: Not applicable.

Data Availability Statement: The original contributions presented in the study are included in the article and Supplementary Material, further inquiries can be directed to the corresponding author/s.

Conflicts of Interest: The authors declare no conflicts of interest.

References

1. Narloch, I.; Wejnerowska, G. An Overview of the Analytical Methods for the Determination of Organic Ultraviolet Filters in Cosmetic Products and Human Samples. *Molecules* **2021**, *26*, 4780. [CrossRef] [PubMed]
2. Wang, A.; Hu, L.; Liu, J.; Tian, M.; Yang, L. Polyaniline-coated core-shell silica microspheres-based dispersive-solid phase extraction for detection of benzophenone-type UV filters in environmental water samples. *Environ. Adv.* **2021**, *3*, 10037. [CrossRef]
3. Barón, E.; Gago-Ferrero, P.; Gorga, M.; Rudolph, I.; Mendoza, G.; Zapata, A.M.; Díaz-Cruz, S.; Barra, R.; Ocampo-Duque, W.; Páez, M.; et al. Occurrence of hydrophobic organic pollutants (BFRs and UV-filters) in sediments from South America. *Chemosphere* **2013**, *92*, 309–316. [CrossRef] [PubMed]
4. Martín-Pozo, L.; del Carmen Gómez-Regalado, M.; Moscoso-Ruiz, I.; Zafra-Gómez, A. Analytical methods for the determination of endocrine disrupting chemicals in cosmetics and personal care products: A review. *Talanta* **2021**, *234*, 122642. [CrossRef] [PubMed]
5. Cadena-Aizaga, M.I.; Montesdeoca-Esponda, S.; Torres-Padrón, M.E.; Sosa-Ferrera, Z.; Santana-Rodríguez, J.J. Organic UV filters in marine environments: An update of analytical methodologies, occurrence and distribution. *Trends Environ. Anal. Chem.* **2020**, *25*, e00079. [CrossRef]
6. Chisvert, A.; Benedé, J.L.; Salvador, A. Current trends on the determination of organic UV filters in environmental water samples based on microextraction techniques—A review. *Anal. Chim. Acta* **2018**, *1034*, 22–38. [CrossRef]
7. Ramos, S.; Homem, V.; Alves, A.; Santos, L. Advances in analytical methods and occurrence of organic UV-filters in the environment—A review. *Sci. Total Environ.* **2015**, *526*, 278–311. [CrossRef]
8. Oubahmane, M.; Mihucz, V.G.; Vasanits, A. Recent trends in the determination of organic UV filters by gas chromatography-mass spectrometry in environmental samples. *TrAC Trends Anal. Chem.* **2023**, *161*, 116995. [CrossRef]
9. Anastas, P.; Eghbali, N. Green Chemistry: Principles and Practice. *Chem. Soc. Rev.* **2010**, *39*, 301–312. [CrossRef]
10. López-Lorente, Á.I.; Pena-Pereira, F.; Pedersen-Bjergaard, S.; Zuin, V.G.; Ozkan, S.A.; Psillakis, E. The ten principles of green sample preparation. *TrAC Trends Anal. Chem.* **2022**, *148*, 116530. [CrossRef]
11. Nowak, P.M.; Wietecha-Posłuszny, R.; Pawliszyn, J. White Analytical Chemistry: An approach to reconcile the principles of Green Analytical Chemistry and functionality. *TrAC Trends Anal. Chem.* **2021**, *138*, 116223. [CrossRef]
12. Keith, L.H.; Gron, L.U.; Young, J.L. Green Analytical Methodologies. *Chem. Rev.* **2007**, *107*, 2695–2708. [CrossRef] [PubMed]
13. Gałuszka, A.; Migaszewski, Z.M.; Konieczka, P.; Namieśnik, J. Analytical Eco-Scale for assessing the greenness of analytical procedures. *TrAC Trends Anal. Chem.* **2012**, *37*, 61–72. [CrossRef]
14. Płotka-Wasylka, J. A new tool for the evaluation of the analytical procedure: Green Analytical Procedure Index. *Talanta* **2018**, *181*, 204–209. [CrossRef] [PubMed]
15. Pena-Pereira, F.; Wojnowski, W.; Tobiszewski, M. AGREE—Analytical GREEnness Metric Approach and Software. *Anal. Chem.* **2020**, *92*, 10076–10082. [CrossRef]
16. Hicks, M.B.; Farrell, W.; Aurigemma, C.; Lehmann, L.; Weisel, L.; Nadeau, K.; Lee, H.; Moraff, C.; Wong, M.; Huangh, Y.; et al. Correction: Making the move towards modernized greener separations: Introduction of the analytical method greenness score (AMGS) calculator. *Green Chem.* **2019**, *21*, 1816–1826. [CrossRef]
17. Manousi, N.; Wojnowski, W.; Płotka-Wasylka, J.; Samanidou, V. Blue applicability grade index (BAGI) and software: A new tool for the evaluation of method practicality. *Green Chem.* **2023**, *25*, 7598–7604. [CrossRef]
18. Płotka-Wasylka, J.; Wojnowski, W. Complementary green analytical procedure index (ComplexGAPI) and software. *Green Chem.* **2021**, *23*, 8657–8665. [CrossRef]
19. Wojnowski, W.; Tobiszewski, M.; Pena-Pereirac, F.; Psillakis, E. AGREEprep—Analytical Greenness Metric for Sample Preparation. *TrAC Trends Anal. Chem.* **2022**, *149*, 116553. [CrossRef]
20. Pena-Pereira, F.; Tobiszewski, M.; Wojnowski, W.; Psillakis, E. A Tutorial on AGREEprep an Analytical Greenness Metric for Sample Preparation. *Advan. Sample Prep.* **2022**, *3*, 100025. [CrossRef]
21. Nowak, P.M.; Kościelniak, P. What Color Is Your Method? Adaptation of the RGB Additive Color Model to Analytical Method Evaluation. *Anal. Chem.* **2019**, *91*, 10343–10352. [CrossRef] [PubMed]
22. Lambropoulou, D.A.; Giokas, D.L.; Sakkas, V.A.; Albanis, T.A.; Karayannis, M.I. Gas chromatographic determination of 2-hydroxy-4-methoxybenzophenone and octyldimethyl-p-aminobenzoic acid sunscreen agents in swimming pool and bathing waters by solid-phase microextraction. *J. Chromatogr. A* **2002**, *967*, 243–253. [CrossRef]
23. Rodil, R.; Moeder, M. Development of a method for the determination of UV filters in water samples using stir bar sorptive extraction and thermal desorption–gas chromatography–mass spectrometry. *J. Chromatogr. A* **2008**, *1179*, 81–88. [CrossRef]
24. Vila, M.; Lamas, J.P.; Garcia-Jares, C.; Dagnac, T.; Llompart, M. Ultrasound-assisted emulsification microextraction followed by gas chromatography–mass spectrometry and gas chromatography–tandem mass spectrometry for the analysis of UV filters in water. *Microchem. J.* **2016**, *124*, 530–539. [CrossRef]

25. Negreira, N.; Rodríguez, I.; Rubí, E.; Cela, R. Dispersive liquid–liquid microextraction followed by gas chromatography–mass spectrometry for the rapid and sensitive determination of UV filters in environmental water samples. *Anal. Bioanal. Chem.* **2010**, *398*, 995–1004. [CrossRef] [PubMed]
26. Negreira, N.; Rodríguez, I.; Ramil, M.; Rubí, E.; Cela, R. Sensitive determination of salicylate and benzophenone type UV filters in water samples using solid-phase microextraction, derivatization and gas chromatography tandem mass spectrometry. *Anal. Chim. Acta* **2009**, *638*, 36–44. [CrossRef]
27. Moeder, M.; Schrader, S.; Winkler, U.; Rodil, R. At-line microextraction by packed sorbent-gas chromatography–mass spectrometry for the determination of UV filter and polycyclic musk compounds in water samples. *J. Chromatogr. A* **2010**, *1217*, 2925–2932. [CrossRef]
28. Wejnerowska, G.; Narloch, I. Determination of Benzophenones in Water and Cosmetics Samples: A Comparison of Solid-Phase Extraction and Microextraction by Packed Sorbent Methods. *Molecules* **2021**, *26*, 6896. [CrossRef]
29. Chung, W.H.; Tzing, S.H.; Ding, W.H. Optimization of dispersive micro solid-phase extraction for the rapid determination of benzophenone-type ultraviolet absorbers in aqueous samples. *J. Chromatogr. A* **2015**, *1411*, 17–22. [CrossRef]
30. Okanouchi, N.; Honda, H.; Ito, R.; Kawaguchi, M.; Saito, K.; Nakazawa, H. Determination of Benzophenones in River-water Samples Using Drop-based Liquid Phase Microextraction Coupled with Gas Chromatography/Mass Spectrometry. *Anal. Sci.* **2008**, *24*, 627–630. [CrossRef]
31. Yılmazcan, Ö.; Kanakaki, C.; Izgi, B.; Rosenberg, E. Fast determination of octinoxate and oxybenzone UV filters in swimming pool waters by gas chromatography/mass spectrometry after solid-phase microextraction. *J. Sep. Sci.* **2015**, *38*, 2286–2297. [CrossRef] [PubMed]
32. Wu, J.W.; Chen, H.C.; Ding, W.H. Ultrasound-assisted dispersive liquid–liquid microextraction plus simultaneous silylation for rapid determination of salicylate and benzophenone-type ultraviolet filters in aqueous samples. *J. Chromatogr. A* **2013**, *1302*, 20–27. [CrossRef] [PubMed]
33. Kawaguchi, M.; Ito, R.; Honda, H.; Endo, N.; Okanouchi, N.; Saito, K.; Saito, K.; Seto, Y.; Nakazawa, H. Simultaneous analysis of benzophenone sunscreen compounds in water sample by stir bar sorptive extraction with in situ derivatization and thermal desorption-gas chromatography-mass spectrometry. *J. Chromatogr. A* **2008**, *1200*, 260–263. [CrossRef] [PubMed]

applied sciences

MDPI

Article

A Survey of UV Filters Used in Sunscreen Cosmetics

Alicja Pniewska and Urszula Kalinowska-Lis *

Department of Cosmetic Raw Materials Chemistry, Faculty of Pharmacy, Medical University of Lodz, 90-419 Lodz, Poland; alicja.pniewska@stud.umed.lodz.pl
* Correspondence: urszula.kalinowska-lis@umed.lodz.pl

Abstract: The aim of this study was to determine the types of UV filters used in adult and children's sunscreen products sold in Poland (part of the EU market) and their frequency of use. The INCI compositions of sunscreen products were collected and analyzed for the presence of UV filters. The study included 150 randomly selected preparations for adults (from 71 brands) and 50 for children (from 33 brands). The survey concerned the UV filters listed in Annex VI to Regulation (EC) No 1223/2009 of the European Parliament and Council of 30 November 2009 on cosmetic products. The most frequently used UV filters in the child sunscreens were triazine derivatives: bis-ethylhexyloxyphenol methoxyphenyl triazine (60.0%) and ethylhexyl triazone (52.0%), and ethylhexyl salicylate (46.0%), a derivative of salicylic acid. The most common in adult sunscreens were butyl methoxydibenzoylmethane (56.0%), a dibenzoylmethane derivative, followed by the salicylic acid derivative ethylhexyl salicylate (54.7%) and the triazine derivatives bis-ethylhexyloxyphenol methoxyphenyl triazine (54.7%) and ethylhexyl triazone (50.0%). Physical filters, including their nano and non-nano forms, were more popular in sunscreens for children, i.e., 50.0% (TiO_2) and 22.0% (ZnO), than for adults: 21.3% (TiO_2) and 6.7% (ZnO). For both adults and children, many cosmetic products contained four or five UV filters per preparation; however, the child preparations often used two UV filters. To summarize, the following UV filters dominate in photoprotectors for both adults and children: butyl methoxydibenzoylmethane, bis-ethylhexyloxyphenol methoxyphenyl triazine, ethylhexyl triazone, ethylhexyl salicylate, and diethylamino hydroxybenzoyl hexyl benzoate.

Keywords: cosmetic regulations; frequency of use; market trends; sunscreen products; UV filters

1. Introduction

UV filters are added to preparations to protect the skin by absorbing or blocking UV light. Sunscreen preparations with UV filters help protect against acute effects of UVR exposure, like tanning, erythema, and immunosuppression, and help prevent phototoxic damage resulting from chronic exposure, such as premature skin aging, pigmentation, collagen degradation, and skin cancer [1–3].

UVA rays (320–400 nm) can penetrate skin deeper than UVB rays, and are responsible for skin carcinogenesis, immunosuppression, hyperpigmentation, and skin aging [3–5]. In turn, UVB rays (290–320 nm) have higher photon energy, and hence can induce potentially mutagenic DNA photoproducts and contribute to the formation of erythema, skin pigmentation, photoimmunosuppression, and skin cancers [6–8].

However, solar radiation is also beneficial for living organisms. It exerts a positive effect on human health, *inter alia* by activating 7-dehydrocholesterol to synthesize Vitamin D in the human skin epidermis or lowering blood pressure through release of nitric oxide. Moreover, exposure to UV rays can improve mood by inducing endorphin release [9,10].

UV filters can be organic agents, often called "chemical", that can absorb UV rays and release thermal energy, or inorganic (mineral) agents, sometimes called "physical", that scatter and reflect UV rays. Organic filters can be divided into the following groups of derivatives: para-aminobenzoic acid esters, salicylic acid derivatives, cinnamic acid derivatives, benzylidenecamphor derivatives, benzophenone derivatives, dibenzoylmethane derivatives, benzimidazole and benzotriazole derivatives, triazine derivatives, and various others, such as Polysilicone-15 [11]. The presence of chromophores with a conjugated system of π- bonds in their molecules is typical for organic filters, most often an aromatic ring bonded to a carbonyl group or connected by a carbon-carbon double bond [11].

The list of UV filters approved for use in cosmetics in the EU (32 entries) is stated in Annex VI to the Cosmetics Regulation (Regulation (EC) No 1223/2009), which also includes their maximum allowed concentrations [12].

Recently, much attention has been paid to the safety of UV filters for humans and the environment. Some UV filters, e.g., avobenzone or ethylhexyl dimethyl PABA, under UV radiation, generate photodegradation products and reactive oxygen species (ROS), causing phototoxicity and/or photoallergic processes in the skin [13–15]. Although some research suggests that homosalate may act as an endocrine disruptor, the SCCS states homosalate is safe for consumers when used in the final product at concentrations up to 7.34% [16]. The SCCS has raised concerns about the endocrine-disrupting properties of 4-methylbenzylidene camphor (4-MBC), including both the thyroid and estrogen systems [17]. The SCCS needs further research to finally determine the safety of benzophenone-3 for the endocrine system, and recommends its use as a UV filter at concentration up to 6% in face cream, hand cream, and lipsticks [18]. In addition, the use of some UV filters, e.g., benzophenone-3 (oxybenzone) or octocrylene, may be associated with adverse effects, including allergic and photoallergic contact dermatitis [18–22]. Photocontact allergy to octocrylene may occur in patients with previous photoallergy to topically applied ketoprofen, but in general, contact allergy attributed to octocrylene appears very rarely [20]. The FDA have highlighted the need for additional safety data on several filters, including octisalate, homosalate, octocrylene, oxybenzone, octinoxate, and avobenzone [23].

The combination of sunscreens with antioxidant and/or anti-inflammatory agents may lower the risk of skin cancer or other skin damage (e.g., sunburn, erythema, inflammation). As such, it has been proposed that some natural products, such as flavonoids, phenolic acids, anthocyanins or carotenoids, or seaweed and plant extracts, may also be used as skin care against UV radiation [24,25].

More recently, contamination from sunscreen products has been found to pose a threat to coastal ecosystems, as they enter the marine environment through direct contact with beachgoers. High concentrations of benzophenone-3 (BP3) and 4-methylbenzylidene camphor (4-MBC), as well as TiO_2 and ZnO, in the surface microlayer were reported in the southern Mediterranean Sea during summer [26–29]. It is believed that sunscreen ingredients may cause bleaching on coral reefs; to counter this, Hawaii, the U.S. Virgin Islands, and Palau took precautionary measures in this regard and withdrew the use of preparations containing benzophenone-3 (BP3, oxybenzone) and ethylhexyl methoxycinnamate (EHMC, octinoxate) [30,31].

The aim of the study was to gain knowledge about the UV filters selected by manufacturers in sunscreen preparations for adults and for children. Learning about the current trends and frequency of use of photoprotective substances on the Polish market (part of the EU market) may be the opportunity to take a closer look at the safety of the most popular UV filters used in sunscreen preparations.

2. Results

The compositions of 150 sunscreen preparations for adults and 50 for children were analyzed for the presence of UV-photoprotective ingredients. Briefly, the labels of the product were searched and the identified products were classified into the appropriate group of derivatives. The analysis also included the number of UV filters per preparation.

2.1. Types of UV Filters and Their Frequency of Use in Preparations

A detailed analysis of the UV filters found in the preparations for children (n = 50) is shown in Figure 1. The most popular substances were two triazine derivatives: bis-ethylhexyloxyphenol methoxyphenyl triazine (BEMT; Tinosorb S), found in 60.0% of the analyzed preparations, and ethylhexyl triazone (EHT), found in 52.0%. These were followed by the salicylic acid derivative—ethylhexyl salicylate (EHS)—present in 46.0%. Finally, the benzophenone derivative diethylamino hydroxybenzoyl hexyl benzoate (DHHB), the dibenzoylmethane derivative butyl methoxydibenzoylmethane (BMBM; Avobenzone), the physical filters titanium dioxide (non-nano) and titanium dioxide (nano), and the cinnamic acid derivative octocrylene (OC) were all found in 20% to 40%.

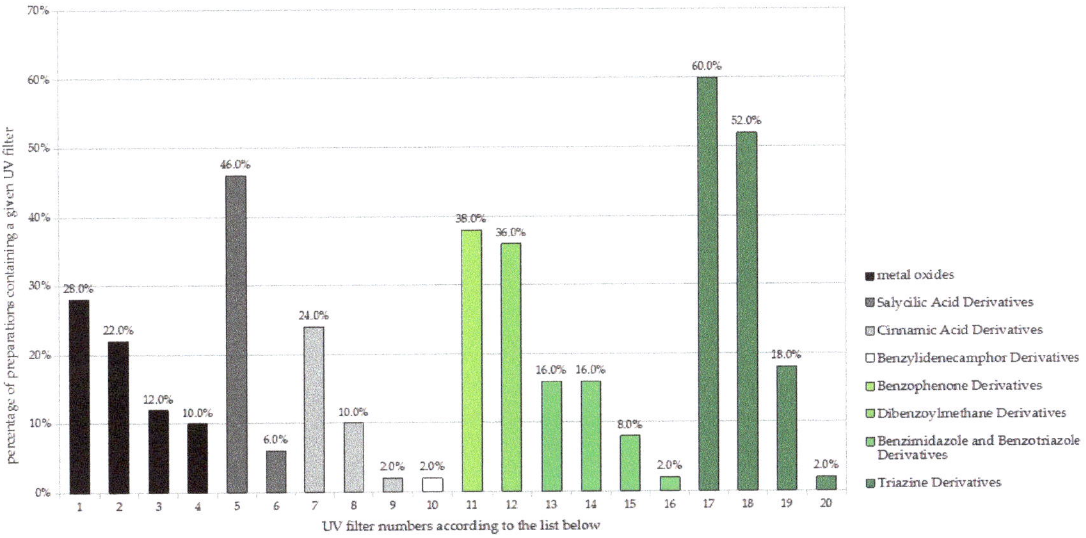

Figure 1. Frequency of use of UV filters in children sunscreen products (n = 50). 1. Titanium dioxide; 2. Titanium dioxide (nano); 3. Zinc oxide; 4. Zinc oxide (nano); 5. Ethylhexyl salicylate (EHS); 6. Homosalate; 7. Octocrylene (OC); 8. Ethylhexyl methoxycinnamate (EHMC); 9. Isoamyl p-methoxycinnamate; 10. Terephthalylidene dicamphor sulfonic acid; 11. Diethylamino hydroksybenzoyl hexyl benzoate (DHHB); 12. Butyl methoxydibenzoylmethane (BMBM; Avobenzone); 13. Phenylbenzimidazole sulfonic acid; 14. Methylene bis-benzotriazolyl tetramethylbutylphenol (nano); 15. Drometrizole trisiloxane; 16. Methylene bis-benzotriazolyl tetramethylbutylphenol; 17. Bis-ethylhexyloxyphenol methoxyphenyl triazine (BEMT; Tinosorb S); 18. Ethylhexyl triazone (EHT); 19. Diethylhexyl butamino triazone; 20. Tris-biphenyl triazine (nano).

Among the adult sunscreens (Figure 2), the largest share was demonstrated by the dibenzoylmethane derivative butyl methoxydibenzoylmethane (BMBM; Avobenzone) with 56.0% of preparations, followed by the salicylic acid derivative ethylhexyl salicylate (EHS) and the triazine derivative bis-ethylhexyloxyphenol methoxyphenyl triazine (BEMT; Tinosorb S), each present in 54.7%, and then a triazine derivative, ethylhexyl triazone (EHT), with 50.0%. The following also had significant percentage shares, ranging from 20% to 40%: diethylamino hydroxybenzoyl hexyl benzoate (DHHB), the cinnamic acid derivatives octocrylene (OC) and ethylhexyl methoxycinnamate (EHMC), and phenylbenzimidazole sulfonic acid (PBSA).

Hence, the most common sunscreen substances in the analyzed cosmetics, both for children and adults, were the organic UV filters from the group of triazine derivatives, salicylic acid derivatives, and dibenzoylmethane derivatives: butyl methoxydibenzoylmethane (BMBM; Avobenzone); bis-ethylhexyloxyphenol methoxyphenyl triazine (BEMT; Tinosorb S);. ethylhexyl salicylate (EHS); ethylhexyl triazone (EHT) (Table 1). Physical filters were more common in preparations for children than for adults. Titanium dioxide (including its nano form) was present in 50.0% of child products and 21.3% of adult products, and zinc oxide (including its nano form) was found in 22.0% of child products but only 6.7% of adult products.

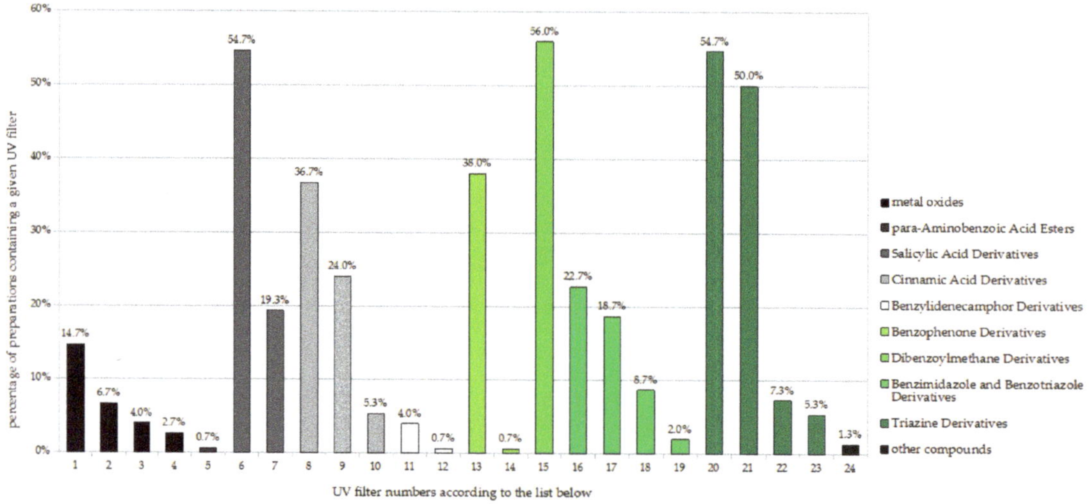

Figure 2. Frequency of use of UV filters in adult sunscreen products (n = 150). 1. Titanium dioxide; 2. Titanium dioxide (nano); 3. Zinc oxide; 4. Zinc oxide (nano); 5. Ethylhexyl dimethyl PABA; 6. Ethylhexyl salicylate (EHS); 7. Homosalate; 8. Octocrylene (OC); 9. Ethylhexyl methoxycinnamate (EHMC); 10. Isoamyl p-methoxycinnamate; 11. Terephthalylidene dicamphor sulfonic acid; 12. 4-Methylbenzylidene camphor; 13. Diethylamino hydroxybenzoyl hexyl benzoate (DHHB); 14. Benzophenone-3; 15. Butyl methoxydibenzoylmethane (BMBM; Avobenzone); 16. Phenylbenzimidazole sulfonic acid (PBSA); 17. Methylene bis-benzotriazolyl tetramethylbutylphenol (nano); 18. Drometrizole trisiloxane; 19. Methylene bis-benzotriazolyl tetramethylbutylphenol; 20. Bis-ethylhexyloxyphenol methoxyphenyl triazine (BEMT; Tinosorb S); 21. Ethylhexyl triazone (EHT); 22. Diethylhexyl butamino tria one; 23. Tris-biphenyl triazine (nano); 24. Polysilicone-15.

Table 1. A profile of the most frequently used UV filters in the analyzed cosmetics.

	INCI Name of a Sunscreen Substance Found in Cosmetics			
Percentage	**for Children**		**for Adults**	
>40%	1. Bis-ethylhexyloxyphenol methoxy-phenyl triazine (BEMT; Tinosorb S) 2. Ethylhexyl triazone (EHT) 3. Ethylhexyl salicylate (EHS)	60.0% 52.0% 46.0%	1. Butyl methoxydibenzoylmethane (BMBM; Avobenzone) 2. Bis-ethylhexyloxyphenol methoxy-phenyl triazine (BEMT; Tinosorb S) 3. Ethylhexyl salicylate (EHS) 4. Ethylhexyl triazone (EHT)	56.0% 54.7% 54.7% 50.0%
20–40%	4. Diethylamino hydroxybenzoyl hexyl benzoate (DHHB) 5. Butyl methoxydibenzoylmethane (BMBM; Avobenzone) 6. Titanium dioxide (TiO_2) 7. Octocrylene (OC) 8. Titanium dioxide (nano) (TiO_2 nano)	38.0% 36.0% 28.0% 24.0% 22.0%	5. Diethylamino hydroxybenzoyl hexyl benzoate (DHHB) 6. Octocrylene (OC) 7. Ethylhexyl methoxycinnamate (EHMC) 8. Phenylbenzimidazole sulfonic acid (PBSA)	38.0% 36.7% 24.0% 22.7%

Some of the identified UV filters were found to be relatively rare, appearing in fewer than 5% of sunscreen products. In the preparations for children, these included isoamyl p-methoxycinnamate, terephthalylidene dicamphor sulfonic acid, methylene bis-benzotriazolyl tetramethylbutylphenol, and tris-biphenyl triazine (nano). For adults, these included zinc oxide, zinc oxide (nano), ethylhexyl dimethyl PABA, terephthalylidene dicamphor sulfonic acid, 4-methylbenzylidene camphor, benzophenone-3, methylen bis-benzotriazolyl tetramethylbutylphenol, and polysilicone-15.

Approximately 30% of the UV filters listed in Annex VI of the Regulation (EC) No 1223/2009 of the European Parliament and of the Council of 30 November 2009 on cosmetic products were not used in any of the tested sunscreen preparations: benzalkonium methosulfate, benzylidene camphor sulfonic acid, polyacrylamidomethyl benzylidene camphor, PEG-25 PABA, benzophenone-4, benzophenone-5, disodium phenyl dibenzimidazole tetrasulfonate, phenylene bis-diphenyltriazine, methoxypropylamino cyclohexenylidene ethoxyethylcyanoacetate, bis-(diethylaminohydroxybenzoyl benzoyl) piperazine, bis-(diethylaminohydroxybenzoyl benzoyl) piperazine (nano), and tris-biphenyl triazine (non-nano form).

2.2. Composition Complexity of the Analyzed Preparations

The complexity of composition, i.e., the number of UV filters per single photoprotective formulation, was recorded. For the sake of brevity, ingredients that do not have a photoprotective function were not included in the list.

The components of the cosmetic products for children are given in Figure 3. Among the 50 analyzed preparations, the largest single group comprised products with two or four components (each 22.0% of all products). In addition, 18.0% of products contained five UV-protective substances in a single preparation, and 14.0% contained six. Products containing three, one, seven, and eight UV filters in one preparation accounted for only 8.0%, 6.0%, 6.0%, and 4.0% of the study group, respectively.

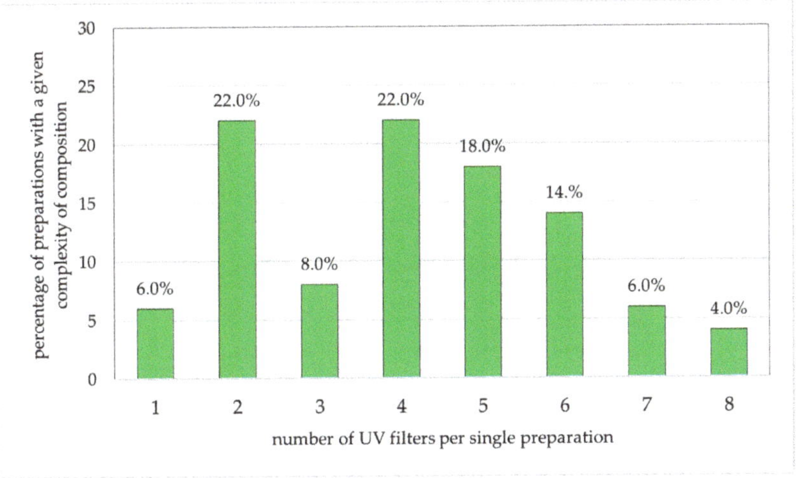

Figure 3. Percentage of preparations for children (n = 50) with specific numbers of sunscreen ingredients (percentages add up to 100%).

Among the adult preparations (Figure 4), the largest single groups of cosmetics included five (27.0%) or four (26.0%) photoprotective substances in a single preparation. Products with three or six components each accounted for 12.7%. The smallest groups included two, one, eight, or nine products, constituting 6.7%, 6.0%, 5.3%, and 0.7%, respectively.

Hence, for both adults and children, the largest group of products include four and five UV-protective substances, while those for children have two components.

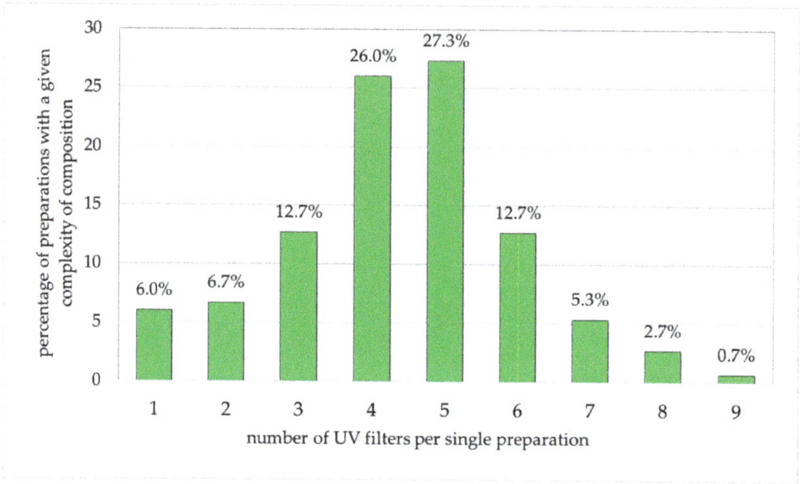

Figure 4. Percentage of preparations for adults (n = 150) with specific numbers of sunscreen ingredients (percentages add up to 100%).

Among the single-component preparations, the most common components were physical UV filters (Table 2). Titanium oxide (nano form) was dominant (100%) in child formulations (n = 3), while TiO_2 (55.6%), ZnO (11.1%), TiO_2 (nano) (11.1%), octocrylene (11.1%), and ethylhexyl methoxycinnamate (11.1%) were found in the adult preparations (n = 9).

Table 2. The preparations containing one or two sunscreen components, with the percentage share of individual substances in a given group.

Analysis of Preparations with One Sunscreen Substance			
for Children		**for Adults**	
Titanium dioxide (nano) (TiO_2 nano)	100%	Titanium dioxide (TiO_2)	55.6%
		Titanium dioxide (nano) (TiO_2 nano)	11.1%
		Zinc oxide (ZnO)	11.1%
		Octocrylene (OC)	11.1%
		Ethylhexyl methoxycinnamate (EHMC)	11.1%
Analysis of preparations with two sunscreen substances			
for children		for adults	
Titanium dioxide (TiO_2)	63.6%	Zinc oxide (ZnO)	40.0%
Zinc oxide (ZnO)	45.5%	Titanium dioxide (TiO_2)	30.0%
Titanium dioxide (nano) (TiO_2 nano)	36.4%	Titanium dioxide (nano) (TiO_2 nano)	20.0%
Zinc oxide (nano) (ZnO nano)	27.3%	Zinc oxide (nano) (ZnO nano)	10.0%
Octocrylene (OC)	18.2%	Phenylbenzimidazole sulfonic acid (PBSA)	20.0%
Bis-ethylhexyloxyphenol methoxyphenyl triazine (BEMT; Tinosorb S)	9.1%	Butyl methoxydibenzoylmethane (BMBM; Avobenzone)	20.0%
		Ethylhexyl salicylate (EHS)	20.0%
		Diethylamino hydroxybenzoyl hexyl benzoate (DHHB)	10.0%
		Octocrylene (OC)	10.0%

The physical UV filters also dominated in the two-component preparations in both study groups (Table 2). In the group of preparations for children (n = 11), the most common were TiO_2 (63.6%), ZnO (45.5%), TiO_2 (nano) (36.4%), and ZnO (nano) (27.3%). Two chemical filters, octocrylene and bis-ethylhexyloxyphenol methoxyphenyl triazine, were present but below the value of 20%. Among those for adults (n = 10), the most prevalent were ZnO (40.0%) and TiO_2 (30.0%), followed by nanometric scale forms of ZnO and TiO_2, and the chemical filters phenylbenzimidazole sulfonic acid, butyl methoxydibenzoylmethane, ethylhexyl salicylate, diethylamino hydroxybenzoyl hexyl benzoate, and octocrylene, all of which were present in 10% to 20% of samples.

Hence, it can be seen that most single- and two-component preparations tended to use physical filters as sunscreen ingredients, and among these, the non-nano form is more common than the nano form.

The complexity of the composition of sunscreen preparations is related to the scope of UV protection of individual filters. When properly selected, the UV filters in a given preparation should fully protect skin against both UVA and UVB radiation. A commonly used combinations of filters that absorb a wide range of UV rays are, e.g., the mixture of Avobenzone (BMBM) with octocrylene (OC) and/or ethylhexyl salicylate (EHS) or the fusion of Avobenzone (BMBM) with Tinosrob S (BEMT), and/or ethylhexyl salicylate (EHS) and/or diethylamino hydroxybenzoyl hexyl benzoate (DHHB). Avobenzone (BMBM) and DHHB provide protection against UVA radiation. Tinosrob S (BEMT) is broad-spectrum filter (UVA and UVB). In turn, the most popular UVB filters include EHT, EHS, OC, EHMC, and PBSA [10,11,19].

2.3. SPF Values of the Analyzed Preparations

The analyzed sunscreen preparations were also reviewed for their SPF value (Figure 5). Among the preparations for children, those with higher SPF values clearly dominated. As many as 82.0% of preparations for children were those with SPF 50 and 14.0% were those with SPF 30. The SPF of the value 20 was the lowest among preparations for children (2.0%). The reviewed adult sunscreens mostly provide high protection of SPF value greater than or equal to 30, with the following distribution: SPF 50 (50.0%), SPF 40 (2.0%), and SPF 30 (30.7%). The remaining adult preparations, with SPF between 6 and 25, accounted for only 9.3%.

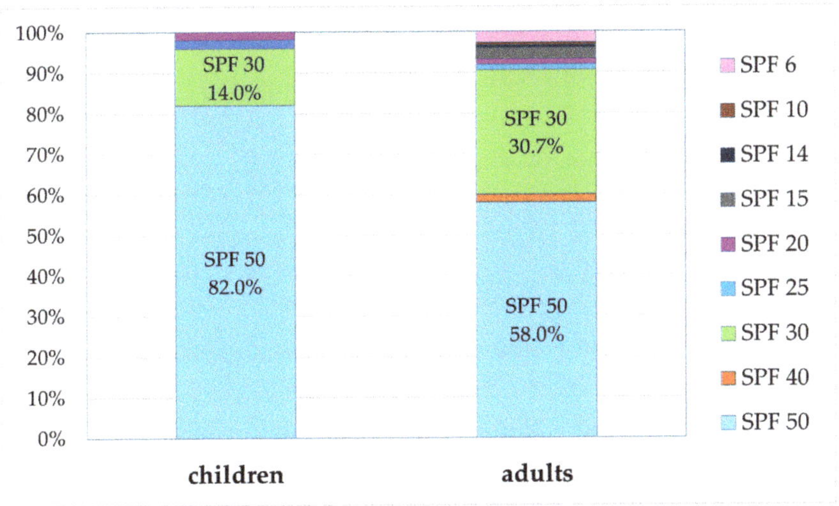

Figure 5. Percentage of preparations for children and adults with a specific SPF value (percentages add up to 100%).

3. Discussion

Our findings indicate that the most popular photoprotective components, both for children and adults, were butyl methoxydibenzoylmethane (BMBM; Avobenzone), bis-ethylhexyloxyphenol methoxyphenyl triazine (BEMT; Tinosorb S), ethylhexyl salicylate (EHS), and ethylhexyl triazone (EHT), and that most sunscreens in Poland have five or four UV filters in a single preparation.

Two surveys of UV filters contained in sunscreen products for children and adults were conducted in Thailand. The first overview (n = 246) was made between December 2017 and March 2018 [32], and the second (n = 312) in April 2020 [33]. While they yielded similar results for the adult products, the reviews of products for children were very divergent; however, this may be due to them being based on small numbers of products, viz. n = 20 [32] and n = 15 [33].

The first overview (n = 226), performed from December 2017 to March 2018 [32], found the most common UV filter for adult to be ethylhexyl methoxycinnamate (EHMC) (62.8%), followed by titanium dioxide, octocrylene (OC), and butyl methoxydibenzoylmethane (BMBM) in 54.9%, 45.1%, and 44.2%, respectively. The most frequently used UV filter for children (n = 20) was butyl methoxydibenzoylmethane (BMBM) (65.0%). The mineral filter TiO_2 (60.0%) was also commonly used, followed by octocrylene (45.0%) and bis-ethylhexyloxyphenol methoxyphenyl triazine (BEMT) (40.0%). Ethylhexyl methoxycinnamate (EHMC) was found less frequently, i.e., only 30.0% [32].

The second survey (n = 297), performed in April 2020 [33], found the most common UV filter for adults to be titanium dioxide (68.0%), followed by ethylhexyl methoxycinnamate (EHMC) (57.2%) and butyl methoxydibenzoylmethane (BMBM) (42.1%), while the most common for children (n = 15) was bis-ethylhexyloxyphenol methoxyphenyl triazine (BEMT) (53.3%), followed by butyl methoxydibenzoylmethane (BMBM) and ethylhexyl salicylate (EHS) (46.7% each). The share for titanium dioxide in this group was 40.0%. [33]. However, these findings cannot be compared with the present study as the filters are intended for people with different skin phototypes from regions with different degrees of sunlight exposure and a different legal environment. Nevertheless, it is worth noting that our results differed meaningfully from the study conducted by Chaiyabutr et al. [33]. The most prevalent filters were 4.5 times (TiO_2) and 2.5 times (EHMC) more prevalent than in our present study; in turn, our predominant filter for adults was butyl methoxydibenzoylmethane (BMBM) (56.0%).

A similar survey of randomly selected preparations from several dozen brands was carried out in Portugal in 2021 [34]. The most popular UV filters for adults (n = 379), i.e., with a percentage share above 35%, included butyl methoxydibenzoylmethane (BMBM) (73.9%), octocrylene (OC) (51.7%), bis-ethylhexyloxyphenol methoxyphenyl triazine (BEMT) (47.5%), ethylhexyl triazone (38.0%), and ethylhexyl salicylate (35.9%). In comparison, in the present study, the prevalence of methoxydibenzoylmethane (BMBM) was 17.9 p.p. lower, with 56.0%; this was followed by bis-ethylhexyloxyphenol methoxyphenyl triazone (BEMT; 54.7%), ethylhexyl salicylate (EHS; 54.7%), and ethylhexyl triazine (EHT; 50.0%). These components also occupied the top positions in the Portuguese study, but with lower shares by 7.2 p.p., 18.8 p.p., and 12.0 p.p., respectively. While octocrylene (OC) was the second most common filter in the Portuguese study (51.7%), it was in sixth place in the present study (36.7%) [34].

Clear differences were also observed regarding child sunscreens between the Polish (n = 50) and Portuguese (n = 65). The predominant component in Poland, bis-ethylhexyloxyphenol methoxyphenyl triazine (BEMT; 60.0%), was only in fifth place in the Portuguese study, with a share of approximately 30%. Conversely, the predominant component in the Portuguese study was butyl methoxydibenzoylmethane (BMBM; approximately 60%), which was only fifth on the list in the present study (36.0%). Both studies had similar prevalence values for other popular filters, i.e., ethylhexyl salicylate (EHS) and ethylhexyl triazone (EHT). In addition, similar values were noted for titanium dioxide (non-nano form) in child formulas in the Portuguese study (35.4%) and the present study (28.0%). Interestingly, the benzophenone derivative DHHB, present in our study in preparations for adults and children with a share of as much as 38%, had, in the Portuguese study, a lower share by approximately 18 p.p. and 13 p.p. [34].

The availability of UV filters varies over time and undoubtedly depends on market forces and legal regulations. A 2014 survey of sunscreen products in the UK (n = 337) found the most common UV filters to be butyl methoxydibenzoylmethane (BMBM) and octocrylene (OC), which were present in 96.4% and 90.5% of sunscreen products. Bis-ethylhexyloxyphenol methoxyphenyl triazine (BEMT) was found in 58.5% of products. Ethylhexyl salicylate (EHT), diethylhexyl butamido triazone, and methylene bis-benzotriazolyl tetramethylbutylphenol were found in around 32% each [35].

In the present study, the most frequently used UVA filter in adult sunscreens was butyl methoxydibenzoylmethane (BMBM; Avobenzone) (56.0%). Its frequency of use in child preparations was 36.0%, putting it in fifth position. Avobenzone is considered one of the most common allergenic and photoallergenic [36], and highly photolabile, UV filters [37]. Photopatch testing has found the reactivity rate of avobenzone to be 1.3–1.7% [38,39]. It loses between 50% and 60% of its protective efficacy after one hour of exposure to UV radiation [40]. Avobenzone undergoes keto-enol tautomerism, and its keto form can easily photodegrade into 4-tert-butyl benzoic acid and 4-methoxy benzoic acid, which are responsible for its photoallergic and phototoxic reactions [11,41].

Photodegradation of unstable UV filters such as avobenzone can be prevented by the use of photostabilizers, encapsulation, antioxidants, and the application of quenchers [42]. In addition, combining avobenzone with other photostable filters such as octocrylene and Tinosorb S can also prevent photodegradation [43,44]. Its photostability can also be improved by loading in cyclodextrin, e.g., beta-cyclodextrin polymers (pbCD) cross-linked by epichlorohydrin (pbCDE), or liposome lipid nanoparticles, microparticles, and polymeric nanoparticles [42,45].

A test of avobenzone and five other organic filters (octisalate, homosalate, octocrylene, oxybenzone, octinoxate) using the ToxCast/Tox21 database found all apart from oxybenzone to have low intrinsic biological activity and a low risk of toxicity, including endocrine disruption, in humans [46].

In the present study, bis-ethylhexyloxyphenol methoxyphenyl triazine (BEMT; Tinosorb S) was the most frequently used UV filter (60.0%) for children and the second most common (54.7%) for adults. Unlike avobenzone, BEMT is photostable. It is a broad absorption spectrum filter (UVB, UVA1, and UVA2) with minimal skin penetration and does not disrupt the functioning of the endocrine system [47,48]. In 1999, SCCS confirmed that there was no evidence that this compound was toxic or allergenic [49].

Ethylhexyl salicylate (EHS) was found to be one of the most popular UV filters in adult and child sunscreens, used in 54.7% and 46.0% of the studied samples. It demonstrates photodegradation and can induce some environmental toxicity [19,34]. Clinical trial data indicates that it is systemically absorbed, resulting in plasma concentrations higher than the FDA systemic exposure threshold (0.5 ng/mL) [50].

Two cinnamic acid derivatives, octocrylene (OC) and ethylhexyl methoxycinnamate (EHMC), were also quite popular UV filters in the present study. They were present in 36.6% (OC) and 24.0% (EHMC) of adult preparations, and 24.0% (OC) and 10.0% (EHMC) of child preparations. Unlike EHMC, octocrylene does not exhibit any endocrine disruption potential nor is it photodegradable [19,20]. OC rarely causes skin irritation reactions (0.6% for n = 1031 [20]. The number of reported cases of allergic contact dermatitis after the use of octocrylene seems to be irrelevant considering the widespread use of it in cosmetic products. Photocontact allergy to octocrylene may occur in patients with previous photoallergy to topically applied ketoprofen [20,51]. Both OC and EHMC demonstrate high bioaccumulation rates, passing into breast milk [19].

Ethylhexyl triazone (EHT), was found to be the second most common UV filter in child sunscreens (52.0%) and the fourth in adults (50.0%). It does not show skin penetration [52] and has good photostability [34]. EHT releases free radicals in contact with sunlight [47].

Titanium dioxide (TiO_2) is an inorganic UV filter present as both nano (22.0%) and micro (28%) forms in child sunscreens. Titanium oxide is considered a more photochemically stable and less skin irritating filter than most organic filters, but it can generate reactive oxygen species (ROS) when exposed to UV radiation, leading to potential adverse effects [53–55]. Therefore, since 2019, titanium dioxide can only be used as a nanomaterial when coated with inert shells like silica, hydrated silica, alumina, aluminium hydroxide, aluminium stearate, stearic acid, trimethoxycaprylylsilane, dimethicone or simethicone, or with some combinations thereof (Annex VI, Regulation No. 1223/2009). TiO_2 (nano) is not allowed in applications that may lead to exposure by inhalation [12].

4. Materials and Methods

From October to December 2023, the INCI compositions of sunscreen products available online on the Polish market were collected. The survey involved 150 randomly chosen preparations with UV filters for adults and 50 for children. The adult products included 71 international brands, with one to eight products from each brand, while the child products were obtained from 33 international brands, with one to three products per brand.

The searched UV filters were listed in Annex VI to Regulation (EC) No 1223/2009 of the European Parliament and Council of 30 November 2009 on cosmetic products.

The collected data were processed using Microsoft Excel. The data collected from the INCI compositions of preparations are series of dichotomies that determine for each preparation and filter whether the filter is included in the preparation or not. There is no numerical measure here, such as concentration or dose. Quantitative analysis of such data encounters great difficulties, so only basic statistical analyses were performed on them.

5. Conclusions

Several popular UV filters are present in sunscreen products on the Polish market, i.e., within the EU. The most common organic UV filters found in child sunscreen products, with a frequency of use above 30%, are bis-ethylhexyloxyphenol methoxyphenyl triazine (BEMT; Tinosorb S)—60.0%, ethylhexyl triazone (EHT)—52.0%, ethylhexyl salicylate (EHS)—46.0%, diethylamino hydroxybenzoyl hexyl benzoate (DHHB)—38.0%, and butyl methoxydibenzoylmethane (BMBM; Avobenzone)—36.0%. The most common adult photoprotectors were butyl methoxydibenzoylmethane (BMBM; Avobenzone)—56.0%, bis-ethylhexyloxyphenol methoxyphenyl triazine (BEMT; Tinosorb S)—54.7%, ethylhexyl salicylate (EHS)—54.7%, ethylhexyl triazone (EHT)—50.0%, diethylamino hydroxybenzoyl hexyl benzoate (DHHB)—38.0%, and octocrylene (OC)—36.7%. The most common organic compounds were triazine, salicylic acid, dibenzoylmethane, and benzophenone derivatives.

Mineral filters, especially titanium dioxide, are much more popular among child sunscreens (TiO_2—28.0% and TiO_2 nano—22.0%) than for adults (TiO_2—14.7% and TiO_2 nano—6.7%). Ten of the UV filters listed in Annex VI to Regulation (EC) No 1223/2009 are not used at all in sunscreen products; these are mostly benzylidenecamphor derivatives, with some benzophenone and triazine derivatives.

Author Contributions: Conceptualization, U.K.-L.; Methodology, A.P.; Formal analysis, A.P.; Investigation, A.P.; Data curation, A.P.; Writing—original draft, A.P.; Writing—review & editing, U.K.-L.; Visualization, A.P.; Supervision, U.K.-L.; Project administration, U.K.-L. All authors have read and agreed to the published version of the manuscript.

Funding: This research was supported by the Medical University of Lodz, grant No. 503/3-066-02/503-31-001.

Institutional Review Board Statement: Not applicable.

Informed Consent Statement: Not applicable.

Data Availability Statement: Data are contained within the article.

Acknowledgments: We would like to thank Elżbieta Budzisz for her skillful contribution and support for the article.

Conflicts of Interest: The authors declare no conflict of interest.

References

1. Seité, S.; Colige, A.; Piquemal-Vivenot, P.; Montastier, C.; Fourtanier, A.; Lapière, A.; Nusgens, B. A full-UV spectrum absorbing daily use cream protects human skin against biological changes occurring in photoaging. *Photodermatol. Photoimmunol. Photomed.* **2000**, *16*, 147–155. [CrossRef] [PubMed]
2. Lautenschlager, S.; Wulf, H.C.; Pittelkow, M.R. Photoprotection. *Lancet* **2007**, *370*, 528–537. [CrossRef] [PubMed]
3. Rittié, L.; Fisher, G.J. Natural and sun-induced aging of human skin. *Cold Spring Harb. Perspect. Med* **2015**, *5*, a015370. [CrossRef] [PubMed]
4. Verkouteren, J.A.C.; Ramdas, K.H.R.; Wakkee, M.; Nijsten, T. Epidemiology of basal cell carcinoma: Scholarly review. *Br. J. Dermatol.* **2017**, *177*, 359–372. [CrossRef] [PubMed]
5. Gibbs, N.K.; Norval, M. Photoimmunosuppression: A brief overview. *Photodermatol. Photoimmunol. Photomed.* **2013**, *29*, 57–64. [CrossRef] [PubMed]
6. Rünger, T.M.; Farahvash, B.; Hatvani, Z.; Rees, A. Comparison of DNA damage responses following equimutagenic doses of UVA and UVB: A less effective cell cycle arrest with UVA may render UVA-induced pyrimidine dimers more mutagenic than UVB-induced ones. *Photochem. Photobiol. Sci.* **2012**, *11*, 207–215. [CrossRef] [PubMed]
7. Bernerd, F.; Asselineau, D. Successive alteration and recovery of epidermal differentiation and morphogenesis after specific UVB-damages in skin reconstructed in vitro. *Dev. Biol.* **1997**, *183*, 123–138. [CrossRef] [PubMed]
8. Lavker, R.M.; Gerberick, G.F.; Veres, D.; Irwin, C.J.; Kaidbey, K.H. Cumulative effects from repeated exposures to suberythemal doses of UVB and UVA in human skin. *J. Am. Acad. Dermatol.* **1995**, *32*, 53–62. [CrossRef] [PubMed]
9. Neale, R.E.; Lucas, R.M.; Byrne, S.N.; Hollestein, L.; Rhodes, L.E.; Yazar, S.; Young, A.R.; Berwick, M.; Ireland, R.A.; Olsen, C.M. The effects of exposure to solar radiation on human health. *Photochem. Photobiol. Sci.* **2023**, *22*, 1011–1047. [CrossRef]
10. Jesus, A.; Sousa, A.; Cruz, M.T.; Cidade, H.; Sousa Lobo, J.M.; Almeida, I.F. UV Filters: Challenges and Prospects. *Pharmaceuticals* **2022**, *15*, 263. [CrossRef]
11. Nitulescu, G.; Lupuliasa, D.; Adam-Dima, I.; Nitulescu, G.M. Ultraviolet Filters for Cosmetic Applications. *Cosmetics* **2023**, *10*, 101. [CrossRef]
12. European Commission. *Regulation (EC) No 1223/2009 of the European Parliament and of the Council of 30 November 2009 on Cosmetic Products*; Official Journal of the European Union L 342/59; Publications Office of the European Union: Luxembourg, 2009.
13. Nash, J.F.; Tanner, P.R. Relevance of UV filter/sunscreen product photostability. *Photodermatol. Photoimmunol. Photomed.* **2014**, *30*, 88–95. [CrossRef] [PubMed]
14. Kawakami, C.M.; Gaspar, L.R. Mangiferin and naringenin affect the photostability and phototoxicity of sunscreens containing avobenzone. *J. Photochem. Photobiol. B Biology* **2015**, *151*, 239–247. [CrossRef] [PubMed]
15. Tarras-Wahlberg, N.; Stenhagen, G.; Larkö, O.; Rosén, A.; Wennberg, A.M.; Wennerström, O. Changes in ultraviolet absorption of sunscreens after ultraviolet irradiation. *J. Investig. Dermatol.* **1999**, *113*, 547–553. [CrossRef] [PubMed]
16. SCCS (Scientific Committee on Consumer Safety). Scientific Advice on the Safety of Homosalate (CAS No 118-56-9, EC No 204-260-8) as a UV-filter in Cosmetic Products in Cosmetic Products; Final Version of 2 December 2021, SCCS/1638/21; European Commission. 2021. Available online: https://inria.hal.science/hal-03464662/ (accessed on 11 April 2024).
17. SCCS (Scientific Committee on Consumer Safety). Scientific Opinion on 4-Methylbenzylidene Camphor (4-MBC); Preliminary Version of 22 December, Final Version of 29 April 2022, SCCS/1640/21; European Commission. 2022. Available online: https://health.ec.europa.eu/document/download/bc7fc1c9-9a5e-4f7c-a67f-a03b4dea312b_en?filename=sccs_o_262.pdf (accessed on 11 April 2024).
18. SCCS (Scientific Committee on Consumer Safety). Opinion on Benzophenone-3 (CAS No 131-57-7, EC No 205-031-5); Preliminary Version of 15 December 2020, Final Version of 30–31 March 2021, SCCS/1625/20; European Commission. 2021. Available online: https://hal.science/hal-03199396/ (accessed on 11 April 2024).
19. Sabzevari, N.; Qiiblawi, S.; Norton, S.A.; Fivenson, D. Sunscreens: UV filters to protect us: Part 1: Changing regulations and choices for optimal sun protection. *Int. J. Women's Dermatol.* **2021**, *7*, 28–44. [CrossRef] [PubMed]
20. Berardesca, E.; Zuberbier, T.; Sanchez Viera, M.; Marinovich, M. Review of the safety of octocrylene used as an ultravioletfilter in cosmetics. *JEADV* **2019**, *33* (Suppl. 7), 25–33. [PubMed]
21. Avenel-Audran, M.; Dutartre, H.; Goossens, A.; Jeanmougin, M.; Comte, C.; Bernier, C.; Benkalfate, L.; Michel, M.; Ferrier-Leбouëdec, M.C.; Vigan, M.; et al. Octocrylene, an emerging photoallergen. *Arch. Dermatol.* **2010**, *146*, 753–757. [CrossRef]
22. Bryden, A.M.; Moseley, H.; Ibbotson, S.H.; Chowdhury, M.M.U.; Beck, M.H.; Bourke, J.; English, J.; Farr, P.; Foulds, I.S.; Gawkrodger, D.J.; et al. Photopatch testing of 1155 patients: Results of the U.K. Multicentre photopatch study group. *Br. J. Dermatol.* **2006**, *155*, 737–747. [CrossRef] [PubMed]
23. FDA (Ed.) Sunscreen Drug Products for Over-the-Counter Human Use. Center for Drug Evaluation and Research, U.S. Department of Health and Human Services. 2019. Available online: https://www.federalregister.gov/documents/2019/02/26/2019-03019/sunscreendrug-products-for-over-the-counter-human-use (accessed on 21 March 2024).

24. Li, L.; Chong, L.; Huang, T.; Ma, Y.; Li, Y.; Ding, H. Natural products and extracts from plants as natural UV filters for sunscreens: A review. *Anim. Models Exp. Med.* **2023**, *6*, 183–195. [CrossRef]
25. Soleimani, S.; Yousefzadi, M.; Babaei Mahani Nezhad, S.; Pozharitskaya, O.N.; Shikov, A.N. Potential of the Ethyl Acetate Fraction of Padina boergesenii as a Natural UV Filter in Sunscreen Cream formulation. *Life* **2023**, *13*, 239. [CrossRef]
26. Thallinger, D.; Labille, J.; Milinkovitch, T.; Boudenne, J.L.; Loosli, F.; Slomberg, D.; Angeletti, B.; Lefrançois, C. UV filter occurrence in beach water of the Mediterranean coast—A field survey over 2 years in Palavas-les-Flots. *Int. J. Cosmet. Sci.* **2023**, *45*, 67–83. [CrossRef] [PubMed]
27. Agawin, N.S.R.; Sunyer-Caldú, A.; Díaz-Cruz, M.S.; Frank-Comas, A.; García-Márquez, M.G.; Tovar-Sánchez, A. Mediterranean seagrass Posidonia oceanica accumulates sunscreen filters. *Mar. Pollut. Bull.* **2022**, *176*, 113417. [CrossRef] [PubMed]
28. Tovar-Sánchez, A.; Sánchez-Quiles, D.; Rodríguez-Romero, A. Massive coastal tourism influx to the Mediterranean Sea: The environmental risk of sunscreens. *Sci. Total Environ.* **2019**, *656*, 316–321. [CrossRef] [PubMed]
29. Sánchez-Quiles, D.; Blasco, J.; Tovar-Sánchez, A. Sunscreen components are a new environmental concern in coastal waters: An overview. In *Sunscreens in Coastal Ecosystems, The Handbook of Environmental Chemistry*; Tovar-Sánchez, A., Sánchez-Quiles, D., Blasco, J., Eds.; Springer: Cham, Switzerland, 2020; Volume 94, pp. 1–14.
30. Miller, I.B.; Pawlowski, S.; Kellermann, M.Y.; Petersen-Thiery, M.; Moeller, M.; Nietzer, S.; Schupp, P.J. Toxic effects of UV filters from sunscreens on coral reefs revisited: Regulatory aspects for "reef safe" products. *Environ. Sci. Eur.* **2021**, *33*, 74. [CrossRef]
31. Heron, S.F.; Maynard, J.A.; Van Hooidonk, R.; Eakin, C.M. Warming trends and bleaching stress of the world's coral reefs 1985–2012. *Sci. Rep.* **2016**, *6*, 38402. [CrossRef] [PubMed]
32. Phadungsaksawasdi, P.; Sirithanabadeekul, P. Ultraviolet filters in sunscreen products labeled for use in children and for sensitive skin. *Pediatr. Dermatol.* **2020**, *37*, 632–636. [CrossRef] [PubMed]
33. Chaiyabutr, C.; Sukakul, T.; Kumpangsin, T.; Bunyavaree, M.; Charoenpipatsin, N.; Wongdama, S.; Boonchai, W. Ultraviolet filters in sunscreens and cosmetic products-A market survey. *Contact Dermat.* **2021**, *85*, 58–68. [CrossRef] [PubMed]
34. Jesus, A.; Augusto, I.; Duarte, J.; Sousa, E.; Cidade, H.; Cruz, M.T.; Lobo, J.M.S.; Almeida, I.F. Recent Trends on UV filters. *Appl. Sci.* **2022**, *12*, 12003. [CrossRef]
35. Kerr, A.C. A survey of the availability of sunscreen filters in the UK. *Clin. Exp. Dermatol.* **2011**, *36*, 541–543. [CrossRef]
36. Heurung, A.R.; Raju, S.I.; Warshaw, E.M. Adverse reactions to sunscreen agents: Epidemiology, responsible irritants and allergens, clinical characteristics, and management. *Dermatitis* **2014**, *25*, 289–326. [CrossRef]
37. Sambandan, D.R.; Ratner, D. Sunscreens: An overview and update. *J. Am. Acad. Dermatol.* **2011**, *64*, 748–758. [CrossRef] [PubMed]
38. Kerr, A.C.; Ferguson, J.; Haylett, A.K.; Rhodes, L.E.; Adamski, H.; Alomar, A.; Serra, E.; Antoniou, C.; Aubin, F.; Vigan, M.; et al. A European multicentre photopatch test study. *Br. J. Dermatol.* **2012**, *166*, 1002–1009.
39. Scalf, L.A.; Davis, M.D.P.; Rohlinger, A.L.; Connolly, S.M. Photopatch testing of 182 patients: A 6-year experience at the Mayo Clinic. *Dermatitis* **2009**, *20*, 44–52. [CrossRef] [PubMed]
40. Bouillon, C. Recent advances in sun protection. *J. Dermatol. Sci.* **2000**, *23* (Suppl. 1), 57–61. [CrossRef]
41. Karlsson, I.; Hillerstrom, L.; Stenfeldt, A.L.; Martensson, J.; Borje, A. Photodegradation of dibenzoylmethanes: Potential cause of photocontact allergy to sunscreens. *Chem. Res. Toxicol* **2009**, *22*, 1881–1892. [CrossRef] [PubMed]
42. Gholap, A.D.; Sayyad, S.F.; Hatvate, N.T.; Dhumal, V.V.; Pardeshi, S.R.; Chavda, V.P.; Vora, L.K. Drug Delivery Strategies for Avobenzone: A Case Study of Photostabilization. *Pharmaceutics* **2023**, *15*, 1008. [CrossRef] [PubMed]
43. Chatelain, E.; Gabard, B. Photostabilization of butyl methoxydibenzoylmethane (Avobenzone) and ethylhexyl methoxycinnamate by bis-ethylhexyloxyphenol methoxyphenyl triazine(Tinosorb S), a new UV broadband filter. *Photochem. Photobiol.* **2001**, *74*, 401–406. [CrossRef] [PubMed]
44. Gaspar, L.R.; Maia Campos, P.M.B.G. Evaluation of the photostability of different UV filter combinations in a sunscreen. *Int. J. Pharm.* **2006**, *307*, 123–128. [CrossRef] [PubMed]
45. Karpkird, T.; Khunsakorn, R.; Noptheeranuphap, C.; Midpanon, S. Inclusion complexes and photostability of UV filters and curcumin with beta-cyclodextrin polymers: Effect on cross-linkers. *J. Incl. Phenom. Macrocycl. Chem.* **2018**, *91*, 37–45. [CrossRef]
46. Onyango, D.O.; Selman, B.G.; Rose, J.L.; Ellison, C.A.; Nash, J.F. Comparison between endocrine activity assessed using ToxCast/Tox21 database and human plasma concentration of sunscreen active ingredients/UV filters. *Toxicol. Sci.* **2023**, *196*, 25–37. [CrossRef]
47. Santander Ballestín, S.; Luesma Bartolomé, M.J. Toxicity of Different Chemical Components in Sun Cream Filters and Their Impact on Human Health: A Review. *Appl. Sci.* **2023**, *13*, 712. [CrossRef]
48. Varrella, S.; Danovaro, R.; Corinaldesi, C. Assessing the eco-compatibility of new generation sunscreen products through a combined microscopic-molecular approach. *Environ. Pollut.* **2022**, *314*, 120212. [CrossRef] [PubMed]
49. Opinion of the Scientific Committee on Cosmetic Products and Non-Food Products Intended for Consumers Concerning 2,4-Bis-{[4-(2-ethyl-hexyloxy)-2-hydroxy]-phenyl}-6-(4-methoxyphenyl)-(1,3,5)-triazine. 17 February 1999. Available online: https://ec.europa.eu/health/archive/ph_risk/committees/sccp/documents/out52_en.pdf (accessed on 20 March 2024).
50. Matta, M.K.; Florian, J.; Zusterzeel, R.; Pilli, N.R.; Patel, V.; Volpe, D.A.; Yang, Y.; Oh, L.; Bashaw, E.; Zineh, I.; et al. Effect of Sunscreen Application on Plasma Concentration of Sunscreen Active Ingredients: A Randomized Clinical Trial. *JAMA* **2020**, *323*, 256–267. [CrossRef] [PubMed]
51. SCCS (Scientific Committee on Consumer Safety). *Opinion on Octocrylene (CAS No 6197-30-4, EC No 228-250-8)*; Preliminary Version of 15 January 2021, Final Version of 30–31 March 2021, SCCS/1627/21; European Commission: Brussels, Belgium, 2021.

52. Souza, C.; Maia Campos, P.M.B.G. Development of a HPLC method for determination of four UV filters in sunscreen and its application to skin penetration studies. *Biomed. Chromatogr.* **2017**, *31*, e4029. [CrossRef] [PubMed]
53. Yuan, S.; Huang, J.; Jiang, X.; Huang, Y.; Zhu, X.; Cai, Z. Environmental Fate and Toxicity of Sunscreen-Derived Inorganic Ultraviolet Filters in Aquatic Environments: A Review. *Nanomaterials* **2022**, *12*, 699. [CrossRef] [PubMed]
54. Schneider, S.L.; Lim, H.W. A review of inorganic UV filters zinc oxide and titanium dioxide. *Photodermatol. Photoimmunol. Photomed.* **2019**, *35*, 442–446. [CrossRef]
55. Bartoszewska, M.; Adamska, E.; Kowalska, A.; Grobelna, B. Novelty Cosmetic Filters Based on Nanomaterials Composed of Titanium Dioxide Nanoparticles. *Molecules* **2023**, *28*, 645. [CrossRef]

applied sciences

MDPI

Article

Efficacy and Tolerability of a Microneedling Device Plus Exosomes for Treating Melasma

Ilaria Proietti [1,†], Chiara Battilotti [1,*,†], Francesca Svara [1], Carlotta Innocenzi [1], Alessandra Spagnoli [2] and Concetta Potenza [1]

1 Dermatology Unit "Daniele Innocenzi", "A. Fiorini" Hospital, Via Firenze, 1, 04019 Terracina, Italy; proiettilaria@gmail.com (I.P.); francescasvara@gmail.com (F.S.); carlottainnocenzi95@gmail.com (C.I.); concetta.potenza@uniroma1.it (C.P.)

2 Department of Public Health and Infectious Diseases, Sapienza University of Rome, 00185 Rome, Italy; alessandra.spagnoli@uniroma1.it

* Correspondence: chiara.battilotti@gmail.com; Tel.: +39-0773-708811

† These authors contributed equally to this work.

Abstract: Melasma is a challenging skin condition which involves both structural and functional skin alterations. Despite the availability of various treatment options, the management remains complex. This is the first study to investigate topical application of *Rosa damascena* stem cell exosomes when used concomitantly with microneedling in women and men with facial melasma. We recruited 20 subjects with Fitzpatrick skin types I-III, exhibiting melasma of varying severity. The modified Melasma Area and Severity Index (mMASI) and Global Aesthetic Improvement Scale (GAIS) were utilized to evaluate treatment response. The treatment protocol involved microneedling followed by exosome application over four or five sessions, at 4-week intervals. Ninety percent of subjects demonstrated a significant improvement in mMASI scores, while only 10% showed no change. GAIS assessment further supports overall improvement, with just 10% categorized as "not changed". Tolerability was favorable, with mild, transient side effects. Our findings suggest promising outcomes with this combined therapy, underscoring its potential as a safe and effective approach for treating melasma, particularly in severe and moderate cases. However, further research with larger sample sizes and control arms is warranted to validate these findings and explore long-term efficacy.

Keywords: melasma; exosomes; microneedling; mMASI; GAIS

1. Introduction

Melasma is a chronic, acquired skin condition characterized by irregular and symmetrical distributed hyperpigmented spots that affects sun-exposed areas, especially the face. It is most commonly found in darker skin types (Fitzpatrick classification III–VI), particularly in Asian women of reproductive age.

1.1. Pathophysiology of Melasma

Melasma is a multifactorial disorder resulting from external factors such as ultraviolet radiation (UVR) exposure, oxidative status, and female hormone stimulation, in genetically predisposed individuals [1]. Wood's lamp examination classifies melasma into epidermal, dermal, or mixed types based on pigment location. However, laser confocal microscopy studies reveal that all melasma types are mixed, suggesting a shared pathophysiology [2]. Although it was initially believed that melasma affected only melanocytes, subsequent histopathological studies of the affected skin have revealed structural and functional alterations across all skin layers [3]. Evidence of solar elastosis, thinning of the stratum corneum (SC), disruption of the basal membrane (BM), and heightened dermal blood vessels suggest that melasma may be a skin condition resulting from "photoaging". Pendulous melanocytes associated with basal membrane abnormality were identified as a distinctive histological feature in melasma [4]. These changes synergistically contribute to hypermelanogenesis [5]. Transcriptomic analyses performed in melasma lesions highlighted the involvement of at least 300 genes, impacting melanocytes and dermal components [6]. Genes involved in melanogenesis, such as tyrosinase (TYR), tyrosinase-related protein 1 (TYRP1), melanocortin 1 receptor (MC1R), and PDZ domain-containing protein 1 (PDZK1), seem to be upregulated. Downregulation of H19 gene impacts melanogenesis and melanin transfer, along with reduced expression of miR-675, which targets microphthalmia-associated transcription factor (MITF), a critical regulator of melanocytic cells. The decreased expression of miR-675 appears to influence cadherin-11 (CDH11), potentially contributing to BM damage [3,5]. UVR and visible light (VL) act as direct stimulants for melanogenesis. Prolonged sun exposure induces fibroblasts to secrete melanogenic and proinflammatory factors, including stem cell factor (SCF), a ligand for the tyrosine kinase receptor c-kit, which contributes to the increase in melanogenesis [7]. UVB has a significant impact on the epidermis and basal membrane; it induces the upregulation of proopiomelanocortin (POMC) in the epidermis, causing melanogenesis and the translocation of melanosomes to keratinocytes. Furthermore, UVB induces the release of inflammatory mediators such as prostaglandins and vascular endothelial growth factor (VEGF), fostering endothelial proliferation and elevating the levels of matrix metalloproteinases, which degrade type IV and VI collagen in the skin. UVA, despite being less erythemogenic, exerts a more significant impact on the superficial dermis through the generation of reactive species, proving to be more efficient in inducing pigment darkening and delayed tanning, especially in individuals with darker phenotypes. Also, VL penetrates the deep dermis and subcutis, but only shorter wavelengths (420–470 nm) can induce pigmentation in darker phototypes by activating opsin 3 (OPN3) receptors in melanocytes [3].

1.2. Available Treatments

Melasma is challenging to treat, and there is currently no known cure. Several treatment options are available, such as chemical peels, lasers, lights, and systemic and topical approaches such as tranexamic acid (TXA), corticosteroids, tretinoin, and hydroquinone, combined with broad-spectrum sunscreen protecting against UVB, long-wave UVA and high-energy VL [8]. Kligman's Trio (KT), initially introduced by Kligman in 1975, is acknowledged as the foremost treatment for melasma. This cream incorporates a blend of 5% hydroquinone, tretinoin, and corticosteroid. Nevertheless, this topical solution is linked to discomforting side effects, such as erythema, desquamation, and a burning sensation. Recently, the use of a new triple combination cream containing isobutylamido thiazolyl resorcinol, retinoic acid, and corticosteroid has been introduced as a well-tolerated alternative to KT [9]. Chemical peels are an effective procedure in treating melasma as they exfoliate superficial skin layers, reducing hyperpigmentation. Various acid substances like salicylic acid, glycolic acid, and trichloroacetic acid, among others, are used at different concentrations to achieve improvement of the disease. They are often used in combination with other treatments, such as laser therapy or topical therapies, to optimize results. The use of deep and medium-depth peels in subjects with darker skin phototypes is not recommended due to the risk of hyperpigmentation [10]. Various types of lasers have become available in melasma treatment, particularly nonablative lasers, being widely utilized due to their lower incidence of post-inflammatory hyperpigmentation. The effectiveness of laser therapy is significantly augmented when combined with complementary topical treatments [11,12].

1.3. Microneedling

Microneedling is a minimally invasive procedure that causes microperforations in the skin using very thin needles. Micro-injuries stimulate neocollagenesis, neo-elastogenesis, and transcutaneous elimination of melanin [13]. Over the past years, this technique, alone or in combination with topical agents, has demonstrated efficacy and safety in melasma treatment, with a growing body of literature supporting its effectiveness. After a single session of microneedling using a dermaroller with 1.5 mm needles, a reduction in melanin density and pendulous melanocyte was observed, along with an improvement in Melasma Area and Severity Index (MASI) and quality of life [14]. Moreover, microneedling improves the effectiveness of topical therapies by increasing their transcutaneous penetration [15]. Using needles of micron size, hydrophilic high molecular weight compounds can penetrate the stratum corneum, which limits transcutaneous absorption of topical creams. This method administers the drug directly into the epidermis or upper dermis layer, ensuring the delivery of 100% of the product [16,17]. The application of topical TXA and a depigmenting solution containing TXA, N-acetyl glucosamine, vitamin C, and idebenone, in conjunction with microneedling, led to further reductions in MASI compared to microneedling alone [18,19]. The benefits of microneedling include a low complication rate and a very low risk of post-inflammatory hyperpigmentation, making it safe even for darker phototypes [20]. Additionally, microneedles do not penetrate deeply enough to reach the pain receptors located in the lower dermis, making the treatment less painful compared to techniques such as hypodermic needles [17].

1.4. Exosomes in Dermatology

Exosomes are lipid bilayer-enclosed nanovesicles released by almost all types of cells carrying proteins, DNA, long non-coding RNAs (lncRNA), micro-RNAs (miRNA), and other bioactive molecules. These nano-sized vesicles typically range in diameter from 30 to 200 nanometers [21] and are formed inside cells in endosomal compartments known as multivesicular bodies (MVBs). When the MVBs fuse with the plasma membrane, they release exosomes into the extracellular space. These exosomes play a key role in intercellular communication and can significantly impact numerous physiological and pathological processes [22]. Exosomes exert local paracrine or distant effects and can be found in various body fluid including plasma, urine, amniotic fluid, and saliva [23]. The cargo of exosomes is dependent on the cell of origin, reflecting its physiological state, type, and environment. It can influence the behavior of recipient cells in several ways, including modulation of gene expression and alteration of cellular processes. The interest in exosomes spans across various fields of biomedical research and clinical application, especially due to their role in signaling mechanism, their potential as biomarkers for disease diagnosis, and their possible role as therapeutic agents and as vehicles for drug delivery. In the field of dermatology and aesthetic medicine, exosomes are emerging as a promising therapeutic option in various conditions, such as hair loss, scar treatment, wound healing, skin aging, and pigmentation disorders [24,25]. In the skin, endogenous exosomes facilitate a complex network of interactions involving keratinocytes, fibroblast, melanocytes, macrophages, adipocytes, and immune cells. They are necessary to maintain cellular functions and tissue homeostasis [26]. Exosomes can influence skin cell behavior, promoting wound healing and collagen synthesis, and modulating melanogenesis. Exosomes carry Wnt proteins to induce Wnt signaling activity in target cells [27]. Wnt signaling is essential for skin development and maintenance, as well as regulation of skin stem cells.

1.4.1. Human Stem Cell-Derived Exosomes

Exogenous exosomes, such as stem cell exosomes, can serve as novel treatment options to repair and rejuvenate skin tissues [28]. Exosomes can be isolated and purified from different sources, such as blood, urine, mesenchymal stem cells, and adipose tissue stem cells. The isolation and characterization of exosomes involve techniques such as ultracentrifugation, size-exclusion chromatography, and flow cytometry. Human umbilical cord mesenchymal stem cell exosomes (hUCMSC-Exos) have been shown to promote wound healing by delivering Wnt4, which activates the Wnt/β-catenin pathway in skin cells. Additionally, they protect skin cells from apoptosis induced by acute heat stress through the activation of AKT pathway [29]. Studies conducted both in vitro and in vivo have highlighted the therapeutic potential of exosomes in treating photodamaged skin by reducing TNF-α levels and increasing TGF-β and tissue inhibitor of MMP (TIMP). This results in an increase in collagen I and elastin, and a reduction in collagen III [30].

Exosomes originating from keratinocytes have been found to enhance melanocyte pigmentation through miR-3196 and MITF-dependent signaling pathways, or via miR-203 and MITF-independent pathways. Conversely, keratinocyte-derived exosomes overexpressing miR-330–5p have been observed to reduce melanin production and TYR expression in melanocytes. Furthermore, miR-675 from keratinocyte exosomes contributes to H19 lncRNA downregulation-stimulated melanogenesis by inhibiting MITF expression [27]. Exosomes derived from human adipose tissue-derived mesenchymal stem/stromal cells (ASC-exosomes) can decrease intracellular melanin content in vitro by affecting downstream factors of TYR (TYRP-1, TYRP-2). However, clinically relevant brightening effects are not evident, suggesting an enhancement in transdermal delivery for more meaningful efficacy [31]. A 12-week split-face study demonstrated that combined treatment with human adipose tissue stem cell-derived exosomes (HASC) and microneedling is effective for facial skin aging, showing improvement in skin hydration, elasticity, and pigmentation. In particular, the melanin index significantly decreased in the skin area treated with exosomes and microneedling compared to the area treated with microneedling alone [32]. Another study explored the therapeutic benefits and the percutaneous penetration of hUCMSC-Exos in conjunction with microneedles, 1565 nm nonablative fractional laser (NAFL), and a plasma device called Peninsula Blue Aurora Shumin Master (PBASM) for the treatment of melasma in both rat models and human subjects. In the animal study, about the effect of penetration, hUCMSC-Exos can penetrate the deep dermis under microneedles, NAFL, and PBASM treatments. All the patients showed significant clinical improvement in melasma compared to baseline, assessed through the MASI score, degree of improvement rate, and physician global assessment score (PGA). No statistically significant differences were found among the three therapeutic approaches [33].

1.4.2. Rose Stem Cell-Derived Exosomes

More recently, it has been demonstrated that rose stem cell exosomes (RSCEs) harbor anti-inflammatory and regenerative properties. Research on plant exosomes is still relatively new. Plant-derived exosomes are similar in structure and function to animal exosomes and they are studied for drug delivery, cancer treatment, inflammatory diseases, and neurodegenerative disorders, offering a novel, cell-free, and sustainable approach to various conditions [34]. Rose stem cells release their exosomes into the conditioned media during callus culture. The size and shape closely resemble exosomes derived from human stem cells. RSCEs are collected by isolating and purifying the supernatant of RSC cultures. Their lipid membrane properties and size, ranging from 30 to 200 nanometers, are verified using Nanoparticle Tracking Analysis (NTA) and Transmission Electron Microscopy (TEM). RSCEs have been shown to enhance collagen production by human dermal fibroblasts by 40–120% in a dose-dependent manner and to increase cellular migration by over 20%. Additionally, RSCEs exhibit anti-inflammatory properties, reducing IL-6 production by macrophages by 50–60%, depending on concentration. Furthermore, RSCEs are internalized by melanocytes, resulting in a decreased melanin content, which indicates a potential whitening effect. The precise molecular underlying this function remains unclear, but it is likely to involve various molecules present within the RSCEs cargo. In summary, RSCEs contain miRNA and peptides with anti-inflammatory properties, promoting fibroblast proliferation, collagen production, and dose-dependent reduction in melanin accumulation [35].

1.5. Aim of the Study

The aim of our study was to evaluate the efficacy and safety of exosomes derived from *Rosa damascena* stem cells, when used concomitantly with microneedling in women and men with facial melasma.

2. Materials and Methods

2.1. Study Design

This monocentric, observational, pilot study was conducted in a private dermato-cosmetology practice and was performed during the autumn and winter season, from September 2023 to January 2024. Twenty subjects (sixteen women and four men) from eighteen to sixty-nine years of age, with Fitzpatrick skin type I–III and facial melasma ranging from mild to severe, were enrolled in this study. Subjects with the following conditions were excluded: skin marks in the experimental area (scars, hypertrichosis, nevi, tattoo), current pregnancy or lactation, history of allergy or reactivity to topical products, and any previous or current topical or systemic therapies for melasma. All the subjects gave their consent before entering the study.

2.2. Study Device

The microneedling device was used with sterile, individually packaged needle cartridges that were disposable. To prevent contamination of the microneedling device, a nonsterile, disposable sheath was employed at the interface between the pen and the needle cartridge (SkinPen® Precision System, Crown Aesthetics, Dallas, TX, USA). The device featured 14 solid needles, each with a diameter of 0.25 mm, operating at a speed ranging from 6300 to 7700 rpm. The maximum extension of the cartridge needles was less than 2.5 mm.

2.3. Intervention

Each subject received microneedling treatment utilizing a dermapen equipped with 1.5 mm long needles. Immediately following the microneedling procedure, a topical application of exosomes (ASCE plus®, Seoul, Republic of Korea) was administered to the treated areas. These combined interventions were performed at 4-week intervals, up to a maximum of five sessions. Only the facial areas affected by melasma were treated. Additionally, all subjects were advised to follow strict sun protection measures throughout the duration of the study.

2.4. Photographic Documentation

Standardized photography was obtained at baseline and during follow-up visits using the VectraH1camera system (Canfield Scientific, Inc., Fairfield, NJ, USA), to achieve more homogeneous and reproducible illumination. Images were taken with standard lighting, with angles set at right lateral 37°, left lateral 37°, frontal, left side and right-side views. Skin analysis is a possible application of VECTRA technology.

2.5. mMASI

mMASI (modified Melasma Area and Severity Index) was scored by dermatologist assessment according to the method developed by Pandya et al. [36]. Total mMASI score range is 0–24 and is categorized as follows: mild (0–8), moderate (9–16), severe (17–24). The following formula was used to calculate the mMASI:

$$\text{mMASI} = 0.3\,(\text{AF})(\text{DF}) + 0.3\,(\text{ALM})(\text{DLM}) + 0.3\,(\text{ARM})(\text{DRM}) + 0.1\,(\text{AC})(\text{DC})$$

where A is proportion of melasma area, D is darkness, F is forehead, LM is the left cheek, RM is the right cheek, and C is the chin.

2.6. GAIS

The Global Aesthetic Improvement Scale was graded by dermatologist assessment as much improved (50% or above), improved (less than 50%), and not improved, regarding the enhancement of the overall cutaneous appearance, in particular skin texture and chromatic homogeneity, as compared with the original condition at baseline before treatment. The elaboration of GAIS also relied on the observation of specific parameters collected by the VECTRA at T0 and T1: spots, wrinkles, texture, and brown spots.

2.7. Evaluation of Tolerability

Tolerability was evaluated by visual inspection of the skin region under examination, performed by a dermatologist. Assessment occurred before treatment and subsequently every 4 weeks before the treatment session until the end of the study. Subjects were encouraged to report any potential discomfort arising from product application (burning, stinging, itching) and any adverse reaction (dryness, erythema, edema, discoloration) along with their severity (mild, moderate, severe).

2.8. Statistical Analysis

Demographic and clinical characteristics are presented as counts and percentages for categorical variables. Comparisons between groups were performed using the chi-square test.

3. Results

3.1. Subjects

Twenty subjects were included in the study and completed it. Population characteristics and number of treatment sessions are listed in Table 1.

Table 1. Subjects baseline characteristic and total number of sessions conducted every 4 weeks.

		Number of Subjects (%)
Age	18–40	2 (10)
	41–50	8 (40)
	51–60	6 (30)
	>60	4 (20)
Sex	Male	4 (20)
	Female	16 (80)
Fitzpatrick Phototype	I	1 (5)
	II	12 (60)
	III	7 (35)
Number of Sessions	4	7 (35)
	5	13 (65)

3.2. mMASI

The mMASI was evaluated at the baseline (T0) and at the end of treatment sessions (T1) (Figure 1). At T0, only one subject exhibited mild mMASI, maintaining this status until the end of the treatment (T1). Among the nine subjects initially classified with moderate MASI at T0, a significant 88.9% transitioned to mild mMASI at T1. Similarly, of the ten subjects with severe mMASI at T0, 40% progressed to mild mMASI (Figure 2), while 60% transitioned to moderate mMASI at T1. Remarkably, 10% of subjects showed no change in mMASI, while 90% of subjects demonstrated improvement.

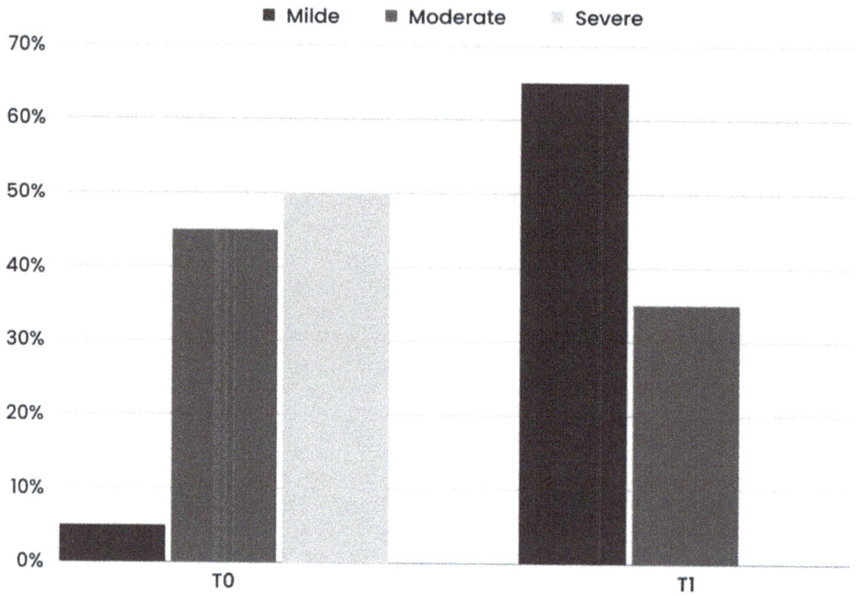

Figure 1. Graphical representation of the percentage of subjects categorized based on mMASI score at T0 and T1.

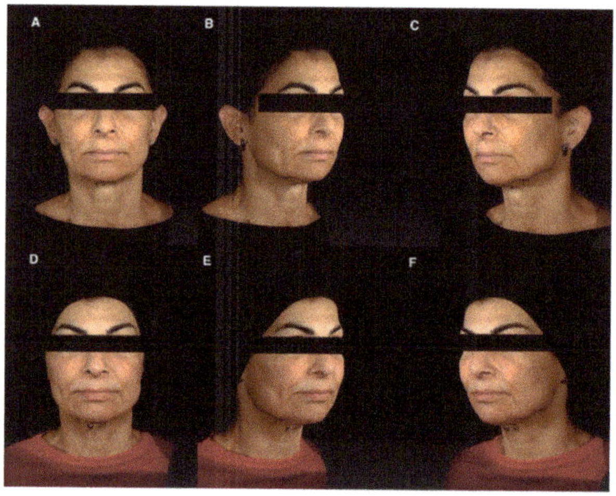

Figure 2. VECTRA® pictures of a patient with severe mMASI at T0 who progressed to mild mMASI at T1. Frontal, right lateral 37°, and left lateral 37° views at baseline (**A–C**); frontal, right lateral 37°, and left lateral 37° views after five sessions of treatment (4 months) (**D–F**).

3.3. GAIS

GAIS was scored by dermatologist assessment at the end of the treatment (T1).

At the end of the treatment sessions, 90% of subjects manifested improvement according to the Global Aesthetic Improvement Scale. In particular, 70% of subjects exhibited an improvement <50% and 20% of subjects showed an improvement >50% compared to the baseline, respectively. A total of 10% of patients showed no improvement (Table 2). Skin parameters obtained with VECTRA acquisitions also confirmed the improvement (Figure 3). No statistically significant associations were found between GAIS and the other variables.

Table 2. Percentage and number of subjects categorized based on GAIS score according to the baseline characteristics and total number of sessions.

		Much Improved 4 (20%)	GAIS Improved 14 (70%)	No Change 2 (10%)
Age (yrs)	18–40	0 (0.0%)	1 (7.1%)	1 (50.0%)
	41–50	3 (75.0%)	5 (35.7%)	0 (0.0%)
	51–60	1 (25.0%)	4 (28.6%)	1 (50.0%)
	>60	0 (0.0%)	4 (28.6%)	0 (0.0%)
Sex	Male	1 (25.0%)	3 (25.0%)	0 (0.0%)
	Female	3 (75.0%)	11 (78.6%)	2 (100%)
Fitzpatrick Phototype	I	0 (0.0%)	1 (7.1%)	0 (0.0%)
	II	4 (100%)	6 (42.9%)	2 (100%)
	III	0 (0.0%)	7 (50.0%)	0 (0.0%)
Number of Sessions	4	1 (25.0%)	4 (28.6%)	2 (100%)
	5	3 (75.0%)	10 (71.4%)	0 (0.0%)

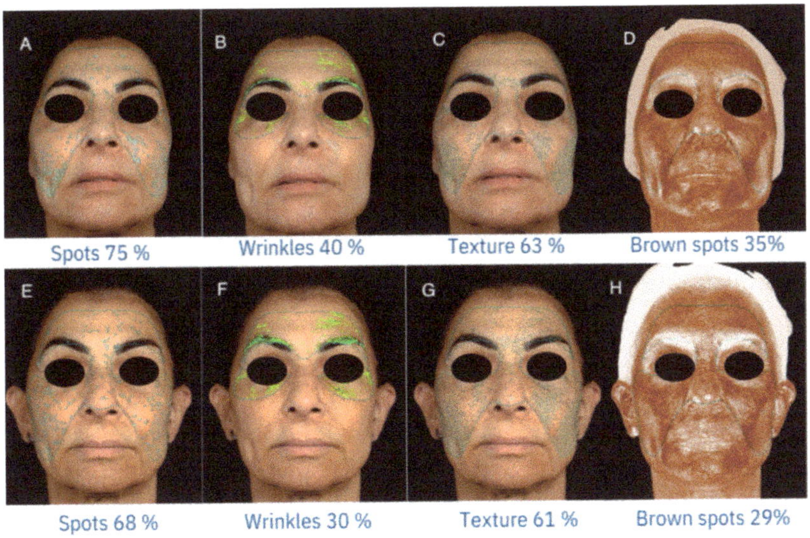

Figure 3. VECTRA evaluation of skin parameters (spots, wrinkles, texture and brown spots) at T0 (**A–D**) and T1 (**E–H**): A reduction in the percentage values related to all four analyzed parameters emerges; in particular, a significant improvement in spots and brown spots is observed.

3.4. Evaluation of Tolerability

All subjects reported mild erythema and burning sensation in the first days following the treatment. No subject reported severe adverse reactions. All subjects reported the treatment as well-tolerated.

4. Discussion

Melasma is a challenging dermatological condition characterized by a tendency for recurrence and difficulties in achieving effective treatment. The condition entails both structural and functional alterations in the epidermis, basement membrane, and upper dermis, resulting in a distinctive hyper melanogenic phenotype. Current therapeutic approaches focus on reducing melanin production and inhibiting melanin synthesis pathways. However, the efficacy of most treatments is limited, and adverse effects are not uncommon [37]. Notably, the epidermal changes associated with melasma go beyond hypermelanosis. Recent histopathological studies have unveiled additional dimensions, exposing solar elastosis, a thinner SC, atrophic granular layer, and modifications in basal keratinocytes [5]. Sun exposure emerges as a pivotal environmental factor in melasma, with both UV and VL playing substantial roles in fostering melanin synthesis, activating tyrosinase activity, and contributing to skin photoaging [7]. The increased vascularity, coupled with heightened VEGF expression, indicates angiogenic involvement in the pathogenesis of melasma. Perivascular mast cells and infiltrating leukocytes contribute to chronic skin inflammation, potentially influencing melanocyte activity and vascular changes [38]. Consequently, an effective therapeutic approach should not only address melanin production but also target the photoaging aspects of the skin.

Originally developed in the 1990s for the treatment of scars, striae, and laxity [39], microneedling has evolved into a well-known procedure for effectively managing melasma. By creating controlled micro-injuries in the epidermis and papillary dermis, a complex local regenerative response is initiated. This encompasses the proliferation of keratinocytes, leading to an increased trans epidermal elimination of melanin, improvement in solar elastosis, and repair of the damaged basal membrane [40]. Furthermore, microneedling enhances the transcutaneous delivery of topical agents, creating microchannels that improves the penetration of active ingredients into the skin strata [41]. In fact, a substantial body of studies have evaluated its combination with topical or oral tranexamic acid, retinoic acid, and vitamin C, yielding encouraging results [10,42]. Due to its swift post-treatment recovery, minimal side effects, its safety in darker skin phototypes [43], and notable clinical outcomes, microneedling stands as a valuable alternative to more invasive approaches like laser skin resurfacing and deep chemical peeling [44].

Stem cell-derived exosomes (SC-Exos) are lipid-bilayer nanovesicles with 30–200 nm molecular diameter. They are produced and secreted from the cells and act as extracellular messengers. SC-Exos share attributes to stem cells (SCs), in addition to the advantages of enhanced stability and low immunogenicity. With antiaging, anti-inflammatory, and antioxidation properties, SC-Exos contribute to skin whitening and promote skin regeneration. In particular, rose stem cell exosomes (RSCEs) promote growth of skin fibroblasts and collagen production, reduced melanin production in melanocytes, and inhibition of inflammation [35]. However, exosomes encounter difficulty penetrating the skin barrier. Recently, it has been demonstrated that microneedling is an effective and safe penetration-promoting method in the treatment of melasma, enhancing the percutaneous penetration of SC-Exos [33].

The presented study underscores the potential efficacy of the combination of *Rosa damascena* stem cell exosomes and microneedling in addressing melasma. It is noteworthy to emphasize that our work benefits from a highly diversified patient sample, representing a strong point of our research. The analysis encompassed both male and female participants, individuals spanning various age groups, and those with different Fitzpatrick skin types. Ninety percent of subjects demonstrated a significant improvement in mMASI scores, while only ten percent showed no change. GAIS assessment further supports overall improvement, with just 10% categorized as "not changed". Tolerability was favorable, marked by mild and transient side effects. It is essential to highlight the variability in responses among subjects, particularly the distinctions between those experiencing better outcomes (transition to mild mMASI) and those with less improvement (transition to moderate mMASI). Notably, half of the subjects not improving in mMASI had a mild form at baseline. This could be attributed to their condition being closer to the desired outcome with fewer pigmentary changes. The totality of subjects with moderate melasma at baseline responded well to treatment, transitioning to mild mMASI. This indicates that this kind of patient may be more receptive to the combined therapy, possibly due to less entrenched melanin deposition or more responsive skin physiology. Fifty percent of the subjects exhibited severe melasma, and all demonstrated a positive response, indicating the efficacy of the combined therapy even in challenging cases. This raises optimism for individuals traditionally deemed to have less favorable prognosis. However, within the severe melasma group, diverse responses were observed. While 40% showed improvement to mild mMASI, 60% transitioned to moderate mMASI, possibly indicating a subgroup less responsive to the intervention, influenced by intrinsic factors or a longer history of melasma. Notably, the absence of severe adverse reactions and post-inflammatory hyperpigmentation, with increased attention to individuals with darker skin types (Fitzpatrick III), along with the perception of the treatment as well-tolerated, indicates a positive safety profile. Finally, the observation that the only subjects not experiencing improvement in GAIS were those who did not undergo the maximum five sessions raises questions about the correlation between the number of sessions and treatment efficacy. This suggests that completing the entire treatment course is crucial for achieving optimal outcomes and discontinuing prematurely might limit the intervention's effectiveness.

5. Conclusions

This is the first study to investigate the clinical treatment with microneedling with topical application of *Rosa damascena* stem cell exosomes. The therapeutic approach outlined in this study has proven to be both safe and effective, requiring no anesthesia and showing no allergic reactions. This makes it a promising option in melasma treatment, especially in patients with severe and moderate melasma at risk of post-inflammatory hyperpigmentation. Limitations of the study include a small sample size and the absence of a control arm with microneedling or topical exosomes application alone. Considering the positive outcome of this pilot study, a randomized control trial with increased sample size could be considered in the future to further evaluate the efficacy of this combined therapeutic approach.

Author Contributions: I.P., C.B. and F.S. participated in the study design and drafting of the manuscript; image assessment, I.P.; data curation, C.I.; statistical analysis, A.S.; supervision, C.P. All authors contributed to the review and final approval of the manuscript. All authors have read and agreed to the published version of the manuscript.

Funding: This research received no external funding.

Institutional Review Board Statement: All subjects gave their informed consent for inclusion before they participated in this study. The study was conducted in accordance with the Declaration of Helsinki, and the protocol was approved by the Ethics Committee of Sapienza University of Rome (Project Identification code: 667).

Informed Consent Statement: All the subjects in this manuscript have provided written informed consent for the publication of their data and photographs.

Data Availability Statement: The datasets created and analyzed during this study are available from the corresponding author upon reasonable request.

Conflicts of Interest: The authors declare no conflicts of interest.

References

1. Zhou, L.L.; Baibergenova, A. Melasma: Systematic review of the systemic treatments. *Int. J. Dermatol.* **2017**, *56*, 902–908. [CrossRef] [PubMed]
2. Kang, W.; Yoon, K.; Lee, E.-S.; Kim, J.; Lee, K.; Yim, H.; Sohn, S.; Im, S. Melasma: Histopathological characteristics in 56 Korean patients. *Br. J. Dermatol.* **2002**, *146*, 228–237. [CrossRef] [PubMed]
3. Espósito, A.C.C.; Cassiano, D.P.; da Silva, C.N.; Lima, P.B.; Dias, J.A.F.; Hassun, K.; Bagatin, E.; Miot, L.D.B.; Miot, H.A. Update on Melasma-Part I: Pathogenesis. *Dermatol. Ther.* **2022**, *12*, 1967–1988. [CrossRef] [PubMed] [PubMed Central]
4. Lee, D.; Park, K.-C.; Ortonne, J.; Kang, H. Pendulous melanocytes: A characteristic feature of melasma and how it may occur. *Br. J. Dermatol.* **2011**, *166*, 684–686. [CrossRef] [PubMed]
5. Kwon, S.-H.; Hwang, Y.-J.; Lee, S.-K.; Park, K.-C. Heterogeneous Pathology of Melasma and Its Clinical Implications. *Int. J. Mol. Sci.* **2016**, *17*, 824. [CrossRef] [PubMed] [PubMed Central]

6. Kang, H.; Hwang, J.; Lee, J.; Ahn, J.; Kim, J.-Y.; Lee, E.-S.; Kang, W. The dermal stem cell factor and c-kit are overexpressed in melasma. *Br. J. Dermatol.* **2006**, *154*, 1094–1099. [CrossRef] [PubMed]
7. Kapoor, R.; Dhatwalia, S.K.; Kumar, R.; Rani, S.; Parsad, D. Emerging role of dermal compartment in skin pigmentation: Comprehensive review. *J. Eur. Acad. Dermatol. Venereol.* **2020**, *34*, 2757–2765. [CrossRef] [PubMed]
8. Rendon, M.; Berneburg, M.; Arellano, I.; Picardo, M. Treatment of melasma. *J. Am. Acad. Dermatol.* **2006**, *54* (Suppl. 2), S272–S281. [CrossRef] [PubMed]
9. Bertold, C.; Fontas, E.; Singh, T.; Gastaut, N.; Ruitort, S.; Pugliese, S.W.; Passeron, T. Efficacy and safety of a novel triple combination cream compared to Kligman's trio for melasma: A 24-week double-blind prospective randomized controlled trial. *J. Eur. Acad. Dermatol. Venereol.* **2023**, *37*, 2601–2607. [CrossRef] [PubMed]
10. Neagu, N.; Conforti, C.; Agozzino, M.; Marangi, G.F.; Morariu, S.H.; Pellacani, G.; Persichetti, P.; Piccolo, D.; Segreto, F.; Zalaudek, I.; et al. Melasma treatment: A systematic review. *J. Dermatol. Treat.* **2022**, *33*, 1816–1837. [CrossRef]
11. Abdel-Raouf Mohamed, H.; Ali Nasif, G.; Saad Abdel-Azim, E.; Abd El-Fatah Ahmed, M. Comparative study of fractional Erbium: YAG laser vs combined therapy with topical steroid as an adjuvant treatment in melasma. *J. Cosmet. Dermatol.* **2019**, *18*, 517–523. [CrossRef]
12. Lee, M.-C.; Chang, C.-S.; Huang, Y.-L.; Chang, S.-L.; Chang, C.-H.; Lin, Y.-F.; Hu, S. Treatment of melasma with mixed parameters of 1,064-nm Q-switched Nd:YAG laser toning and an enhanced effect of ultrasonic application of vitamin C: A split-face study. *Lasers Med. Sci.* **2014**, *30*, 159–163. [CrossRef] [PubMed]
13. Jung, J.W.; Kim, W.O.; Jung, H.R.; Kim, S.A.; Ryoo, Y.W. A face-split study to evaluate the effects of microneedle radiofrequency with Q-switched Nd:YAG laser for the treatment of melasma. *Ann. Dermatol.* **2019**, *31*, 133–138. [CrossRef] [PubMed]
14. Cassiano, D.; Espósito, A.C.; Hassun, K.; Lima, E.d.A.; Bagatin, E.; Miot, H. Early clinical and histological changes induced by microneedling in facial melasma: A pilot study. *Indian J. Dermatol. Venereol. Leprol.* **2019**, *85*, 638–641. [CrossRef]
15. Yadav, S.; Singh, A. Microneedling: Advances and widening horizons. *Indian Dermatol. Online J.* **2016**, *7*, 244–254. [CrossRef]
16. Moro, F.; Camela, E.; Samela, T.; Pirrotta, L.; Pupa, M.B.; Zingoni, T.; Fusco, I.; Colonna, L. 1064 nm Q-Switched Fractional Laser for Transcutaneous Delivery of a Biostimulator: Efficacy and Safety Outcomes of a Split-Face Study. *Cosmetics* **2024**, *11*, 14. [CrossRef]
17. Waghule, T.; Singhvi, G.; Dubey, S.K.; Pandey, M.M.; Gupta, G.; Singh, M.; Dua, K. Microneedles: A smart approach and increasing potential for transdermal drug delivery system. *Biomed. Pharmacother.* **2018**, *109*, 1249–1258. [CrossRef] [PubMed]
18. Saleh, F.Y.; Abdel-Azim, E.S.; Ragaie, M.H.; Guendy, M.G. Topical tranexamic acid with microneedling versus microneedlingalone in treatment of melasma: Clinical, histopathologic, and immunohistochemical study. *J. Egypt. Women's Dermatol. Soc.* **2019**, *16*, 89–96. [CrossRef]
19. Farshi, S.; Mansouri, P. Study of efficacy of microneedling and mesoneedling in the treatment of epidermal melasma: A pilot trial. *J. Cosmet. Dermatol.* **2020**, *19*, 1093–1098. [CrossRef]
20. Cohen, B.E.; Elbuluk, N. Microneedling in skin of color: A review of uses and efficacy. *J. Am. Acad. Dermatol.* **2015**, *74*, 348–355. [CrossRef]
21. Ha, D.H.; Kim, H.-K.; Lee, J.; Kwon, H.H.; Park, G.-H.; Yang, S.H.; Jung, J.Y.; Choi, H.; Lee, J.H.; Sung, S.; et al. Mesenchymal Stem/Stromal Cell-Derived Exosomes for Immunomodulatory Therapeutics and Skin Regeneration. *Cells* **2020**, *9*, 1157. [CrossRef] [PubMed] [PubMed Central]
22. American Association of Neurological Surgeons (AANS); American Society of Neuroradiology (ASNR); Cardiovascular and Interventional Radiology Society of Europe (CIRSE); Canadian Interventional Radiology Association (CIRA); Congress of Neurological Surgeons (CNS); European Society of Minimally Invasive Neurological Therapy (ESMINT); European Society of Neuroradiology (ESNR); European Stroke Organization (ESO); Society for Cardiovascular Angiography and Interventions (SCAI); Society of Interventional Radiology (SIR); et al. Multisociety Consensus Quality Improvement Revised Consensus Statement for Endovascular Therapy of Acute Ischemic Stroke. *Int. J. Stroke* **2018**, *13*, 612–632. [CrossRef] [PubMed]
23. Bano, R.; Ahmad, F.; Mohsin, M. A perspective on the isolation and characterization of extracellular vesicles from different biofluids. *RSC Adv.* **2021**, *11*, 19598–19615. [CrossRef] [PubMed] [PubMed Central]
24. Thakur, A.; Shah, D.; Rai, D.; Parra, D.C.; Pathikonda, S.; Kurilova, S.; Cili, A. Therapeutic Values of Exosomes in Cosmetics, Skin Care, Tissue Regeneration, and Dermatological Diseases. *Cosmetics* **2023**, *10*, 65. [CrossRef]
25. Olumesi, K.R.; Goldberg, D.J. A review of exosomes and their application in cutaneous medical aesthetics. *J. Cosmet. Dermatol.* **2023**, *22*, 2628–2634. [CrossRef] [PubMed]
26. Yin, L.; Liu, X.; Shi, Y.; Ocansey, D.K.W.; Hu, Y.; Li, X.; Zhang, C.; Xu, W.; Qian, H. Therapeutic Advances of Stem Cell-Derived Extracellular Vesicles in Regenerative Medicine. *Cells* **2020**, *9*, 707. [CrossRef]
27. Gross, J.C.; Chaudhary, V.; Bartscherer, K.; Boutros, M. Active Wnt proteins are secreted on exosomes. *Nat. Cell Biol.* **2012**, *14*, 1036–1045. [CrossRef] [PubMed]
28. Xiong, M.; Zhang, Q.; Hu, W.; Zhao, C.; Lv, W.; Yi, Y.; Wang, Y.; Tang, H.; Wu, M.; Wu, Y. The novel mechanisms and applications of exosomes in dermatology and cutaneous medical aesthetics. *Pharmacol. Res.* **2021**, *166*, 105490. [CrossRef] [PubMed]
29. Zhang, B.; Wu, X.; Zhang, X.; Sun, Y.; Yan, Y.; Shi, H.; Zhu, Y.; Wu, L.; Pan, Z.; Zhu, W.; et al. Human umbilical cord mesenchymal stem cell exosomes enhance angiogenesis through the Wnt4/β-catenin pathway. *Stem Cells Transl. Med.* **2015**, *4*, 513–522. [CrossRef] [PubMed] [PubMed Central]

30. Ku, Y.C.; Omer Sulaiman, H.; Anderson, S.R.; Abtahi, A.R. The Potential Role of Exosomes in Aesthetic Plastic Surgery: A Review of Current Literature. *Plast. Reconstr. Surg. Glob. Open* **2023**, *11*, e5051. [CrossRef] [PubMed] [PubMed Central]
31. Cho, B.S.; Lee, J.; Won, Y.; Duncan, D.I.; Jin, R.C.; Lee, J.; Kwon, H.H.; Park, G.-H.; Yang, S.H.; Park, B.C.; et al. Skin brightening efficacy of exosomes derived from human adipose tissue-derived stem/stromal cells: A prospective, split-face, randomized placebo-controlled study. *Cosmetics* **2020**, *7*, 90. [CrossRef]
32. Park, G.H.; Kwon, H.H.; Seok, J.; Yang, S.H.; Lee, J.; Park, B.C.; Shin, E.; Park, K.Y. Efficacy of combined treatment with human adipose tissue stem cell-derived exosome-containing solution and microneedling for facial skin aging: A 12-week prospective, randomized, split-face study. *J. Cosmet. Dermatol.* **2023**, *22*, 3418–3426. [CrossRef] [PubMed]
33. Wang, T.; Gao, H.; Wang, D.; Zhang, C.; Hu, K.; Zhang, H.; Lin, J.; Chen, X. Stem cell-derived exosomes in the treatment of melasma and its percutaneous penetration. *Lasers Surg. Med.* **2022**, *55*, 178–189. [CrossRef] [PubMed]
34. Mu, N.; Li, J.; Zeng, L.; You, J.; Li, R.; Qin, A.; Liu, X.; Yan, F.; Zhou, Z. Plant-Derived Exosome-Like Nanovesicles: Current Progress and Prospects. *Int. J. Nanomed.* **2023**, *18*, 4987–5009. [CrossRef] [PubMed] [PubMed Central]
35. Yu Jin, W.; Esther, L.; Seon Young, M.; Byong Seung, C. Biological Function of Exosome-like Particles Isolated from Rose (Rosa Damascena) Stem Cell Culture Supernatant. *bioRxiv* **2023**, 2023.10.17.562840. [CrossRef]
36. Pandya, A.G.; Hynan, L.S.; Bhore, R.; Riley, F.C.; Guevara, I.L.; Grimes, P.; Nordlund, J.J.; Rendon, M.; Taylor, S.; Gottschalk, R.W.; et al. Reliability assessment and validation of the Melasma Area and Severity Index (MASI) and a new modified MASI scoring method. *Am. Acad. Dermatol.* **2011**, *64*, 78–83.e2. [CrossRef]
37. Ma, W.; Gao, Q.; Liu, J.; Zhong, X.; Xu, T.; Wu, Q.; Cheng, Z.; Luo, N.; Hao, P. Efficacy and safety of laser-related therapy for melasma: A systematic review and network meta-analysis. *J. Cosmet. Dermatol.* **2023**, *22*, 2910–2924. [CrossRef] [PubMed]
38. Zhu, J.W.; Ni, Y.J.; Tong, X.Y.; Guo, X.; Wu, X.P. Activation of VEGF receptors in response to UVB promotes cell proliferation and melanogenesis of normal human melanocytes. *Exp. Cell Res.* **2020**, *387*, 111798. [CrossRef] [PubMed]
39. Orentreich, D.S.; Orentreich, N. Subcutaneous Incisionless (Subcision) Surgery for the Correction of Depressed Scars and Wrinkles. *Dermatol. Surg.* **1995**, *21*, 543–549. [CrossRef]
40. Cassiano, D.P.; Espósito, A.C.C.; Hassun, K.M.; de Andrade Lima, M.M.D.; de Andrade Lima, E.V.; Miot, L.D.B.; Miot, H.A.; Bagatin, E. Histological changes in facial melasma after treatment with triple combination cream with or without oral tranexamic acid and/or microneedling: A randomised clinical trial. *Indian J. Dermatol. Venereol. Leprol.* **2022**, *88*, 761–770. [CrossRef]
41. Hou, A.; Cohen, B.; Haimovic, A.; Elbuluk, N. Microneedling: A Comprehensive Review. *Dermatol. Surg.* **2017**, *43*, 321–339. [CrossRef] [PubMed]
42. Brasil dos Santos, J.; Nagem Lopes, L.P.; de Lima, G.G.; Teixeira da Silva, R.; da Silva e Souza Lorca, B.; Miranda Pinheiro, G.; Faria de Freitas, Z.M. Microneedling with cutaneous delivery of topical agents for the treatment of melasma: A systematic review. *J. Cosmet. Dermatol.* **2022**, *21*, 5680–5695. [CrossRef] [PubMed]
43. Rivas, S.; Pandya, A.G. Treatment of melasma with topical agents, peels and lasers: An evidence-based review. *Am. J. Clin. Dermatol.* **2013**, *14*, 359–376. [CrossRef]
44. Alster, T.S.; Graham, P.M. Microneedling: A review and practical guide. *Dermatol. Surg.* **2018**, *44*, 397–404. [CrossRef] [PubMed]

applied sciences

MDPI

Article

Efficacy and Safety Assessment of Topical Omega Fatty Acid Product in Experimental Model of Irritant Contact Dermatitis: Randomized Controlled Trial

Magdalena Ivic [1], Ana Slugan [1], Dario Leskur [1], Doris Rusic [1], Ana Seselja Perisin [1], Darko Modun [1], Toni Durdov [1], Josko Bozic [2], Dubravka Vukovic [3] and Josipa Bukic [1,*]

[1] Department of Pharmacy, University of Split School of Medicine, 21000 Split, Croatia; magdalena.ivic16@gmail.com (M.I.); ana.slugan2@gmail.com (A.S.); dleskur@mefst.hr (D.L.); drusic@mefst.hr (D.R.); aperisin@mefst.hr (A.S.P.); dmodun@mefst.hr (D.M.); toni.durdov@mefst.hr (T.D.)
[2] Department of Pathophysiology, University of Split School of Medicine, 21000 Split, Croatia; jbozic@mefst.hr
[3] Department of Dermatovenerology, University Hospital Split, 21000 Split, Croatia; dvukovic@mefst.hr
* Correspondence: jbukic@mefst.hr

Abstract: Contact dermatitis is a common inflammatory skin disease that often requires prescription therapy and is associated with adverse reactions. Omega fatty acids have been recognized for their anti-inflammatory effect and could serve as a safer option in contact dermatitis treatment. Therefore, the aim of this randomized controlled study, conducted at the University of Split School of Medicine, was to evaluate the efficacy and safety of omega fatty acids containing topical products in an experimental model of irritant contact dermatitis. This study was registered with ClinicalTrials (NCT06189144) and is closed. The primary outcomes were levels of transepidermal water loss, skin hydration, and skin erythema, all measured using an MPA6 device in 25 healthy participants. A significant difference was observed between the hydration values of the intervention (45.7 ± 12.4) and control groups (31.6 ± 12.3) ($p < 0.05$) on final measurements (day 10). Moreover, higher erythema levels were observed in participants who were smokers, compared to non-smokers. No adverse drug reactions were observed during the study period. In conclusion, omega fatty acids topical product use shows promise in the treatment of irritant contact dermatitis, and further studies are needed to evaluate efficacy in a larger sample of patients.

Keywords: contact dermatitis; eczema; skin barrier; omega acids; Rilastil; irritant contact dermatitis; randomized controlled trial; transepidermal water loss; erythema; skin hydration

1. Introduction

Contact dermatitis, an inflammatory skin disease, is characterized by red lesions of the skin which are itchy. These lesions are caused by contact with some external substance. Contact dermatitis, and inflammatory skin diseases in general, significantly impact the affected individuals' quality of life [1,2]. Moreover, these diseases could cause work-related disabilities, financial costs, and social embarrassment. One of the inflammatory skin diseases commonly seen in clinical practice is irritant contact dermatitis. The prevalence of this disease is highly variable worldwide, and this type of dermatitis is mainly caused by the exposure of the skin to an irritant substance, after which damage to the epidermis function occurs. The outer epidermis of the skin serves multiple protective functions, as it controls the permeability of the skin, secretes antimicrobial peptides, includes antioxidants in its extracellular matrix, and secures the physiological hydration levels of the skin [3,4].

In the past, it was believed that irritant contact dermatitis was triggered by direct toxicity of the irritant substance without activation of the immune system. However, later studies have shown that immune-system response is associated with irritant contact dermatitis, and penetration of irritants leads to inflammatory cytokine production by keratinocytes. This production is followed by the recruitment of immune cells (mast cells, macrophages, etc.) [5]. Therefore, the pharmacotherapy of contact dermatitis includes medicines that target lymphocytes, interleukins, and cytokines. Examples of these medicines are delgocitinib, ruxolitinib, dupilumab, tralokinumab, and small molecule AFX5931. However, all of these drugs are registered for systemic use, which is accompanied by the risk of adverse drug reactions and drug interactions. Some of the reported adverse drug reactions include hypersensitivity reactions, respiratory reactions, and ocular adverse reactions. However, as the majority of these medicines have been marketed in the past few years, the safety profiles will be evaluated in the future, according to adverse drug reaction reports from patients and health care professionals [2].

Even though many therapeutic agents have been evaluated for the treatment of inflammatory skin diseases and their exacerbation, glucocorticoids remain a gold standard in their therapy, and immunomodulatory medications are another option; however, their adverse reactions prevent their wide utilization, and it is necessary to find an effective replacement in clinical practice, e.g., therapies for inflammatory skin disease with a better safety profile [6]. Moisturizing agents have been recognized as an adjuvant therapy in patients with not only contact dermatitis, but also psoriasis and atopic dermatitis. Conversely, not all of the available moisturizers have a role in the improvement of skin barrier function or the effect on the skin's susceptibility to various irritant stimuli [7,8].

Omega polyunsaturated fatty acids have been widely used for their antioxidant and anti-inflammatory effects, mainly as an oral supplement recommended for individuals with chronic non-communicable diseases, such as cardiovascular diseases [9]. However, recent studies shed light on omega use in skin diseases, especially in inflammatory skin diseases such as acne [10], atopic dermatitis [11], and psoriasis [12]. The majority of these studies examined the efficacy of oral use of omega fatty acids, and research on the topical use of products containing omega acids is sparse. Current results suggest that omega-3 fatty acids decrease the production of cytokines involved in inflammation (e.g., interleukin 6) and stimulate Treg cells, which produce anti-inflammatory factors. The stimulation of anti-inflammatory factors in the skin is a proposed mechanism of action for omega-6 fatty acids as well [13].

Therefore, the aim of our study was to evaluate the efficacy and safety of an omega-containing topical formulation in the model of irritant contact dermatitis.

2. Materials and Methods

Our study was designed as a randomized controlled clinical trial. The study setting was the University of Split, School of Medicine, Department of Pharmacy, and it was conducted in June 2023. The Ethics Committee of the University of Split School of Medicine approved the research, and the study was conducted in accordance with ethical principles.

In total, 25 healthy male and female volunteers aged 21 to 29 participated in the research. The subjects had no previously recorded skin diseases. Every participant signed an informed consent before the start of the study. The informed consent form contained information about the aim of the research, the method, and all procedures of its implementation. The exclusion criteria were the following: skin diseases, skin cancers, sun damage in the place intended for examination, use of immunomodulatory medicine, corticosteroids, or antihistamines 30 days before the start of the study, application of emollient cream 3 days before the start of the study, non-compliance with the protocol, excessive exposure to natural or artificial ultraviolet radiation, pregnancy and lactation, history of vitiligo, melasma and other hyperpigmentation or photosensitivity disorders, immunosuppression, and allergy to some of the ingredients of the cream used for the intervention. Before inclusion in the study, participants were also asked if they had ever experienced a side effect from any of the ingredients of the researched cream.

The subjects' forearms were chosen as the test site. On each of the forearms, the place was selected where the irritant dermatitis model would be utilized using a 1% (m/m) sodium lauryl sulfate (SLS) solution. After the baseline skin parameters were measured, a volume of 60 μL of the solution of SLS was applied to a piece of paper and kept under occlusion for twenty-four hours under twelve-millimeter Finn chambers [14]. Participants were reminded that they should avoid contact with moisture within the above-mentioned period. After removing the Finn's chamber, the test site was washed with water, and the skin parameters were measured again.

After skin damage was inflicted, a computer program in Excel was used to randomly select which forearm would receive the treatment and which would remain untreated. The treatment used was Rilastil Difesa Crema Sterile (Istituto Ganassini, Milan, Italy), which contains omega-3, omega-6, and omega-9 fatty acids. Additionally, 5 mg/cm^2 of the cream was applied twice daily to the selected forearm with at least an eight-hour interval between applications. On measurement days, the first application of the cream was performed by the examiners immediately after measuring skin parameters. The second application, as well as all applications on non-measurement days, was performed by the subjects themselves according to the given instructions. The last application of the cream before measurement had to be at least twelve hours prior to measurement to ensure the accuracy of the Transepidermal Water Loss (TEWL) measurement. All subjects received one tube of the cream. The cream was weighed before the start and after the completion of the study to confirm adherence to the protocol.

Three skin parameters were measured using a non-invasive bioengineering method with the MPA6 device (Courage + Khazaka GmBH, Cologne, Germany), as these parameters were recognized as fundamental for skin barrier assessment. TEWL is an objective measure of the skin barrier, quantified by the amount of water that evaporates through the skin to the external environment [15]. In our study, skin barrier function was determined by measuring TEWL with the Tewameter TM 300 probe (Courage + Khazaka GmBH, Cologne, Germany). The Tewameter® TM 300 probe indirectly measures the water diffusion from the stratum corneum by using two pairs of spatially separated sensors (a relative humidity sensor and a thermometer) that are located inside the hollow cylinder. This open-chamber method measures the evaporation rate and expresses the values in g/m^2 h [16].

The Corneometer CM 825 (Courage + Khazaka GmBH, Cologne, Germany) was used to measure skin hydration. Corneometer® is based on capacitance measurements and is well known for its precision and sensitivity in skin hydration measurements [17,18]. The instrument converts the skin's capacitance values into arbitrary units of skin hydration, which range from 0 to 120 [19]. Skin erythema was measured using the Mexameter MX 18 probe (Courage + Khazaka GmBH, Cologne, Germany). The base of this measurement is the absorption and reflection of light from the skin. A total of 16 light-emitting diodes (LEDs), which are circularly placed on the top of the instrument, emit light at three wavelengths: 568, 660, and 870 nm, which correspond to green, red, and infrared light, respectively. Light that is reflected by the skin is measured by a photo-detector, and the amount of light absorbed by the skin can be calculated. Green and red light are used in erythema measurements to match the hemoglobin's spectral absorption peak [20]. All measurements were conducted in accordance with the guidelines of the Standardization Group of the European Society of Contact Dermatitis [21].

After the objective measurement of skin parameters with the probes, participants were asked about their subjective impressions of using the cream and any potential side effects such as itching, irritation, and similar symptoms. The sites were also examined by a board-certified dermatologist during the study period.

The total duration of the study was eleven days, during which seven measurements were taken. On the first day, the baseline values of the skin parameters were measured, after which a Finn chamber with SLS was applied to the skin. On the second day, after removing the chamber, the values of skin damage were measured, and the intervention was introduced. On the third, fourth, fifth, eighth, and eleventh days, the recovery values of the skin on both arms were measured. On other days of the study, therapy was applied without additional measurements.

Statistical analysis was conducted using IBM SPSS Statistics software (version 25). Two-way ANOVA with repeated measures was used to compare changes over time between the two groups. An Independent Samples t-Test was utilized to compare the means of the two groups at each specific time point. A Paired Samples t-Test was used to compare values within the same group at the beginning and the end of the study. Statistical significance was set at $p < 0.05$. Data are presented as mean ± standard deviation unless stated otherwise.

3. Results

3.1. Demographic Characteristics

In the study, all twenty-five participants were fully measured through all measurement points or days of the study. A greater proportion of the participants in this study were female (80% of all participants). The largest proportion of participants were 24 years old, and all included participants were between 22 and 28 years old (the median age was 24). A greater proportion of participants reported being non-smokers compared to those who used various tobacco products (60% vs. 40%).

3.2. TEWL

The baseline TEWL values were compared between the intervention group and the control group. There was no difference between the baseline values in the control and intervention groups. As shown in Figure 1, TEWL peaked after irritation of the skin with SLS in both the control and intervention groups (41.7 ± 17.1 vs. 42.2 ± 14.6 g m^{-2} h^{-1}, control and intervention). In subsequent measurements, TEWL values decreased until day 7 when they were approximated to the baseline values (19.4 ± 3.7 vs. 18.4 ± 4.83 g m^{-2} h^{-1}, control and intervention). There was no significant difference in magnitude and speed of normalizing TEWL between the groups even with the use of the cream.

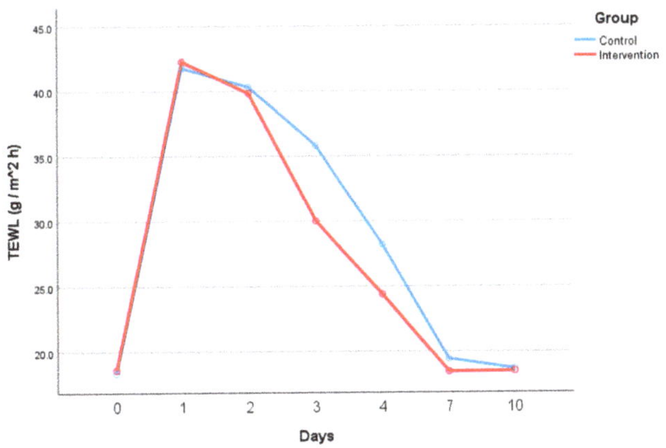

Figure 1. TEWL over the study period represented as two separate curves. One for the control group and a second for the intervention group.

Changes in TEWL over time were also compared between smokers and non-smokers for both the intervention and control groups. Baseline values for smokers and non-smokers were similar for both the control (18.0 ± 6.8 vs. 19.0 ± 5.8 g m^{-2} h^{-1}, non-smokers and smokers) and intervention groups (18.6 ± 4.9 vs. 18.5 ± 4.3 g m^{-2} h^{-1}, non-smokers and smokers). There was no significant difference in TEWL between smokers and non-smokers, neither in the control nor in the intervention group. TEWL values were comparable for both groups for all measurements made. In this study, smoking did not interfere with TEWL recovery after SLS irritation.

3.3. Hydration

Hydration was also compared between the control and intervention groups, and also over time. As seen in Figure 2, both groups had similar baseline values (42.9 ± 8.7 vs. 42.0 ± 6.6 arbitrary units, control and intervention), and both groups experienced a sharp hydration-value decrease after SLS irritation (36.4 ± 11.1 vs. 35.6 ± 11.6 arbitrary units, control and intervention). While the control group's hydration values recovered to the baseline values, the group that used the cream had its hydration values increase over baseline values. Moreover, the use of the cream was shown to keep skin hydrated for a longer period, while the control group had its hydration values decrease again after some time. On day 10, there was a significant difference between the hydration values of the intervention (45.7 ± 12.4 arbitrary units) and control groups (31.6 ± 12.3 arbitrary units) ($p < 0.05$). Respondents' skin was more hydrated when using the cream, compared to not using it.

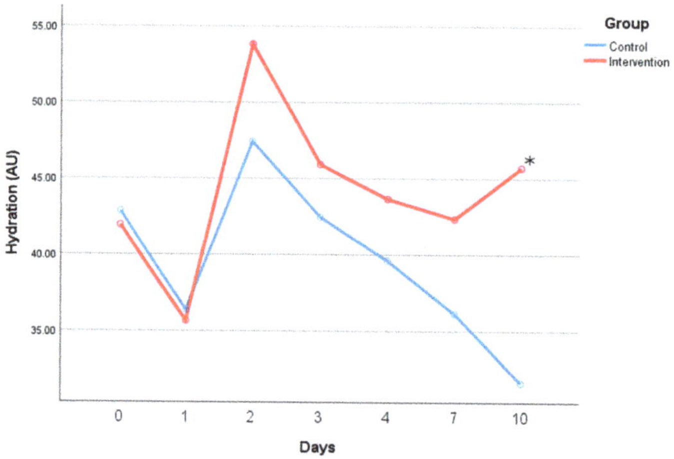

Figure 2. Hydration values over the study period. Comparing control and intervention groups. * Paired Samples *t*-Test.

Hydration values between smokers and non-smokers were also compared for both groups. Figure 3 shows that baseline values for smokers were slightly higher than those values for non-smokers in the control group, but values changed with the same dynamic with no differences based on using tobacco products.

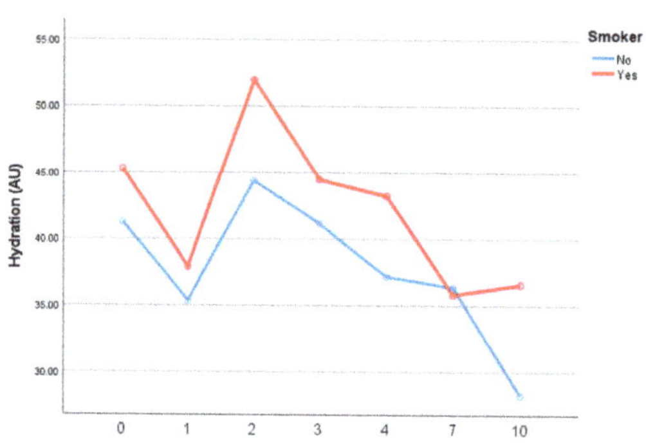

Figure 3. Hydration values over the study period. Comparison between smokers and non-smokers in the control group.

The same comparison was conducted in the intervention group and the result was similar, as seen in Figure 4. There was no significant difference between the hydration values of smokers and non-smokers for the period of the study. Surprisingly, smokers had higher hydration values at day 10, but the difference was not significant (43.8 ± 10.9 vs. 48.6 ± 14.5 arbitrary units, control and intervention). In this study, smoking did not interfere with the hydration values of either group.

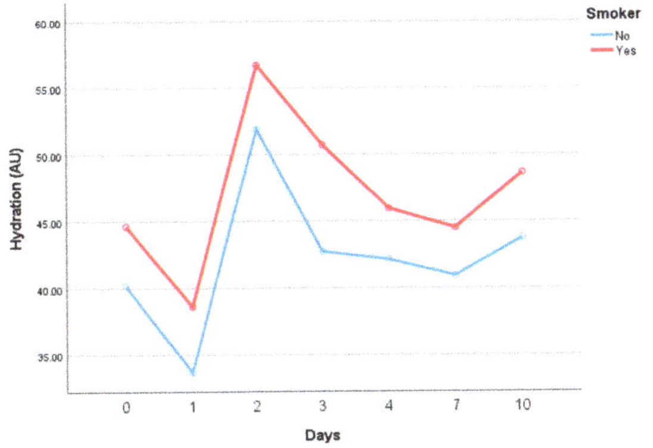

Figure 4. Hydration values over the study period. Comparison between smokers and non-smokers in the intervention group.

3.4. Erythema

Erythema values were measured for 10 days for both the control and intervention groups. There was no significant difference in baseline values between the groups. Values in both groups increased after irritation with SLS, but erythema was more present in the intervention group, compared to the control group, on day 3, as we see in Figure 5 (236.9 ± 58.1 vs. 268.3 ± 51.4 arbitrary units, control and intervention). It took more time for the intervention group to return to baseline erythema values, although there was no significant difference between the groups in any measurement.

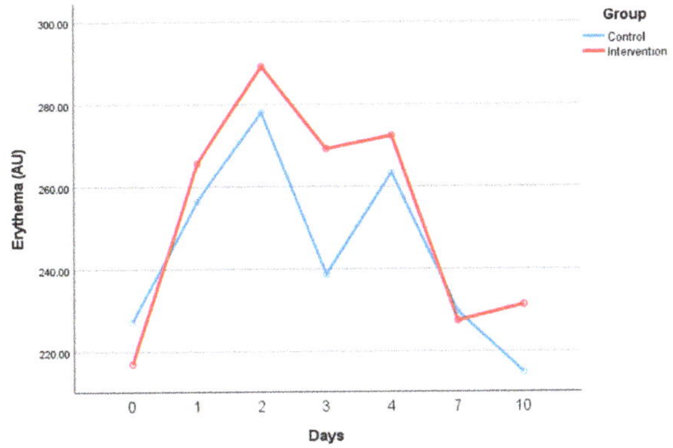

Figure 5. Erythema values over the study period. Comparison between control and intervention groups.

As with previous parameters, erythema values were compared based on the tobacco-using habits of the respondents. Figure 6 illustrates a similar pattern of erythema value changes in both smokers and non-smokers in the control group. There is no significant difference between the groups. However, there was a clear difference between smokers' and non-smokers' erythema presence in the intervention group on day 7 of the study (207.2 ± 47.5 vs. 251.6 ± 50.5 arbitrary units, control and intervention). Smokers had higher erythema values while using the cream than non-smokers using the cream, as seen in Figure 7 ($p < 0.05$).

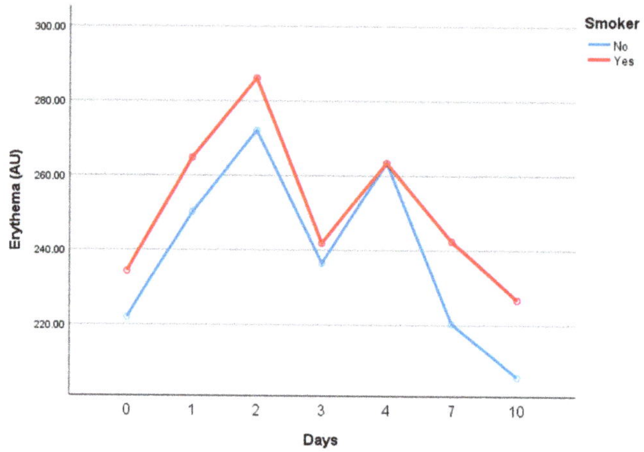

Figure 6. Erythema values over the study period. Comparison between smokers and non-smokers in the control group.

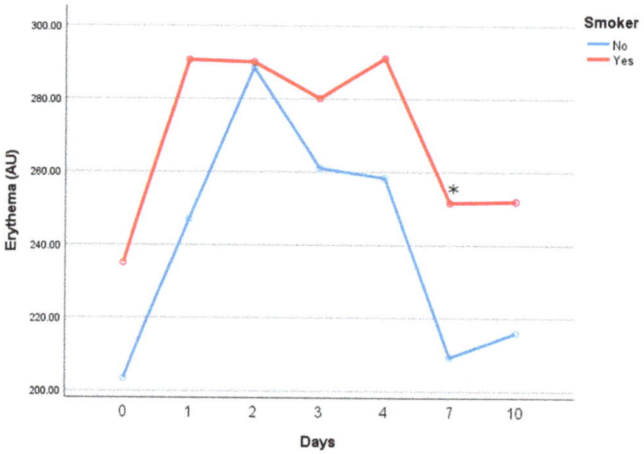

Figure 7. Erythema values over the study period. Comparison between smokers and non-smokers in the intervention group. * Paired Samples *t*-Test.

4. Discussion

To our knowledge, this is one of the first studies examining the efficacy and safety of omega fatty acid topical products in irritant contact dermatitis. After SLS exposure, both the intervention and control groups showed increased TEWL values, decreased hydration levels, and increased erythema values, all of which suggest that the skin barrier function has been impaired. During the study period, TEWL and erythema measurements returned to their baseline values in both groups. However, hydration levels were significantly higher in the group where the Rilastil cream was applied, compared to the control group. Moreover, during the study period, an increase in hydration levels was observed in the intervention group, compared to baseline values. Proper hydration levels of the skin could help maintain skin homeostasis, which is important in individuals affected with inflammatory skin diseases, but also for the general population [22]. Decreased levels of hydration could affect the ability of the skin to combat free radical-induced oxidative stress and inflammation, all of which could lead to inflammation of the skin, accompanied by the risk of infections and other skin complications. However, additional research is needed to better understand the risk factors that affect inflammation [23–27].

Previous studies on patients with inflammatory skin diseases also reported the positive effect of omega fatty acids on skin hydration levels [13]. For instance, a study by Young Park et al. examined the effect of oral use of primrose oil, which contains omega acid, in acne patients treated with isotretinoin [28]. Isotretinoin is a prescription-only medication used in acne therapy, and it is associated with skin dryness as a common adverse reaction. The results of this study showed an increase in hydration levels in the group of patients who used the oil, compared to the control group. Similar results were found in a study by Brosche et al. on patients with atopic dermatitis [29]. During the study period, participants did not report any adverse reaction to the product, which suggests its acceptable safety profile in the treatment of contact dermatitis. Furthermore, another advantage of this product is its route of application, since its topical therapy is the first-line treatment compared to systemic medications, especially in pediatric patients [30].

Bioactive products such as omega fatty acids, polyphenols, carotenoids, etc., gained popularity in the maintenance of healthy skin [31]. Previous studies have shown that a deficiency of fatty acids results in skin dryness, inflammation, and an increased susceptibility to environmental irritants [32]. One of the most promising omega fatty acids, gamma-linolenic acid, when topically applied, seems to penetrate the skin and maintain the health of the stratum corneum; when taken orally, it enhances the cohesiveness of the dermis and prevents excessive TEWL [33]. Since the results of our study did not show a significant difference between the TEWL values of the intervention and control groups, future studies should include data on circulating omega fatty acid levels in individuals. The hypothesis is that topical treatment would show greater improvement in TEWL measurements in individuals with lower levels of circulating omega fatty acids, and this hypothesis should be further evaluated in future randomized controlled clinical trials. Indeed, a similar observation was found in a study by Marchlewicz et al. where patients with severe psoriasis had a significantly lower concentration of omega-3 and omega-9 compared to patients with less severe psoriasis [34].

The results of our study also showed higher values of erythema in participants who were smokers, both in the intervention and control groups. Smoking is well recognized as a risk factor for skin aging, but the consensus on the effect of smoking on skin barriers has not been reached [35,36]. The results of a cross-sectional study by Alotaibi et al. showed a strong association between smoking and irritant contact dermatitis [37]. Recently, Hergesell et al. examined changes in skin barrier proteins and lipids, and their results showed perturbation of the barrier proteins and lipids, even in the skin areas that were not directly exposed to cigarette smoking [38]. It should be noted that no differences in baseline measurements between smokers and non-smokers were observed in our study. Therefore, future studies should involve a larger sample size of both smokers and non-smokers in order to further evaluate the impact of smoking on skin parameters such as TEWL, hydration levels, and erythema.

This study has some limitations. First of all, it included a small sample size. Nevertheless, since the cream containing omega showed promising results, it seems reasonable to examine it in a larger sample size in the future. Secondly, the intervention cream was applied twice a day, and participants applied the cream on their own; however, to neutralize this possible bias, one application per measurement day was performed under our supervision. Additionally, our study was conducted at one geographic location, with no direct comparison with different population samples. Furthermore, ethnicity could also be a potential variability factor, as it has long been proven that the skin of different ethnic populations can have different compositions and characteristics.

5. Conclusions

The use of omega fatty acid cream in irritant contact dermatitis participants significantly increases the hydration values of the skin, but these results should be confirmed in a larger sample size. Moreover, as skin hydration is essential in all inflammatory skin diseases, further studies could include patients with atopic dermatitis and psoriasis.

Author Contributions: Conceptualization, D.R.; Data Curation, A.S., T.D. and D.V.; Formal Analysis, T.D.; Investigation, M.I. and A.S.; Methodology, D.L., D.V. and J.B. (Josipa Bukic); Software, A.S.P.; Supervision, D.M.; Validation, J.B. (Josko Bozic); Visualization, A.S.P.; Writing—Original Draft, M.I. and J.B. (Josipa Bukic); Writing—Review and Editing, D.L., D.R., D.M. and J.B. (Josko Bozic). All authors have read and agreed to the published version of the manuscript.

Funding: This research received no external funding.

Institutional Review Board Statement: This study was conducted in accordance with the Declaration of Helsinki, and approved by the Ethics Committee of the University of Split School of Medicine (protocol code 2181-98-03-04-23-0030, 23 April 2023).

Informed Consent Statement: Informed consent was obtained from all subjects involved in the study.

Data Availability Statement: Data is available from the corresponding author upon a reasonable request.

Acknowledgments: We are grateful to Istituto Ganassini, Milan, Italy, for a donation of Difesa creams.

Conflicts of Interest: The authors declare no conflicts of interest.

Abbreviations

The following abbreviations are used in this manuscript:

SLS	Sodium lauryl sulfate
TEWL	Transepidermal water loss

References

1. Sheikh, H.M.; Jha, R.K. Triggered Skin Sensitivity: Understanding Contact Dermatitis. *Cureus* **2024**, *16*, 59486. [CrossRef] [PubMed]
2. Tancredi, V.; Buononato, D.; Caccavale, S.; Di Brizzi, E.V.; Di Caprio, R.; Argenziano, G.; Balato, A. New Perspectives in the Management of Chronic Hand Eczema: Lessons from Pathogenesis. *Int. J. Mol. Sci.* **2024**, *25*, 362. [CrossRef] [PubMed]
3. Pesqué, D.; Aerts, O.; Bizjak, M.; Gonçalo, M.; Dugonik, A.; Simon, D.; Ljubojević-Hadzavdić, S.; Malinauskiene, L.; Wilkinson, M.; Czarnecka-Operacz, M.; et al. Differential diagnosis of contact dermatitis: A practical-approach review by the EADV Task Force on contact dermatitis. *J. Eur. Acad. Dermatol. Venereol.* **2024**. [CrossRef] [PubMed]
4. Elias, P.M. Skin barrier function. *Curr. Allergy Asthma Rep.* **2008**, *8*, 299–305. [CrossRef] [PubMed]
5. Malekpour, M.; Etebari, A.; Hezarosi, M.; Anissian, A.; Karimi, F. Mouse Model of Irritant Contact Dermatitis. *Iran. J. Pharm. Res.* **2022**, *21*, e130881. [CrossRef] [PubMed]
6. Kang, S.Y.; Um, J.Y.; Chung, B.Y.; Lee, S.Y.; Park, J.S.; Kim, J.C.; Park, C.W.; Kim, H.O. Moisturizer in Patients with Inflammatory Skin Diseases. *Medicina* **2022**, *58*, 888. [CrossRef] [PubMed]
7. Lodén, M. Barrier recovery and influence of irritant stimuli in skin treated with a moisturizing cream. *Contact Dermat.* **1997**, *36*, 256–260. [CrossRef] [PubMed]
8. Lodén, M. Effect of moisturizers on epidermal barrier function. *Clin. Dermatol.* **2012**, *30*, 286–296. [CrossRef] [PubMed]
9. Djuricic, I.; Calder, P.C. Beneficial Outcomes of Omega-6 and Omega-3 Polyunsaturated Fatty Acids on Human Health: An Update for 2021. *Nutrients* **2021**, *13*, 2421. [CrossRef]
10. Jung, J.Y.; Kwon, H.H.; Hong, J.S.; Yoon, J.Y.; Park, M.S.; Jang, M.Y.; Suh, D.H. Effect of dietary supplementation with omega-3 fatty acid and gamma-linolenic acid on acne vulgaris: A randomised, double-blind, controlled trial. *Acta Derm. Venereol.* **2014**, *94*, 521–525. [CrossRef]
11. Kawamura, A.; Ooyama, K.; Kojima, K.; Kachi, H.; Abe, T.; Amano, K.; Aoyama, T. Dietary supplementation of gamma-linolenic acid improves skin parameters in subjects with dry skin and mild atopic dermatitis. *J. Oleo Sci.* **2011**, *60*, 597–607. [CrossRef] [PubMed]
12. Balbás, G.M.; Regaña, M.S.; Millet, P.U. Study on the use of omega-3 fatty acids as a therapeutic supplement in treatment of psoriasis. *Clin. Cosmet. Investig. Dermatol.* **2011**, *4*, 73–77. [CrossRef] [PubMed]

13. Balić, A.; Vlašić, D.; Žužul, K.; Marinović, B.; Bukvić Mokos, Z. Omega-3 Versus Omega-6 Polyunsaturated Fatty Acids in the Prevention and Treatment of Inflammatory Skin Diseases. *Int. J. Mol. Sci.* **2020**, *21*, 741. [CrossRef] [PubMed]
14. Casari, A.; Farnetani, F.; De Pace, B.; Losi, A.; Pittet, J.; Pellacani, G.; Longo, C. In Vivo Assessment of Cytological Changes by Means of Reflectance Confocal Microscopy—Demonstration of the Effect of Topical Vitamin E on Skin Irritation Caused by Sodium Lauryl Sulfate. *Contact Dermat.* **2017**, *76*, 131–137. [CrossRef] [PubMed]
15. Green, M.; Kashetsky, N.; Feschuk, A.; Maibach, H.I. Transepidermal water loss (TEWL): Environment and pollution-A systematic review. *Skin Health Dis.* **2022**, *2*, 104. [CrossRef] [PubMed]
16. Fluhr, J.; Wiora, G.; Nikolaeva, D.G.; Miséry, L.; Darlenski, R. In vivo transepidermal water loss: Validation of a new multi-sensor open chamber water evaporation system Tewameter TM Hex. *Skin Res. Technol.* **2023**, *29*, 13307. [CrossRef]
17. Westermann, T.V.A.; Viana, V.R.; Berto Junior, C.; Detoni da Silva, C.B.; Carvalho, E.L.S.; Pupe, C.G. Measurement of skin hydration with a portable device (SkinUp® Beauty Device) and comparison with the Corneometer®. *Skin Res. Technol.* **2020**, *26*, 571–576. [CrossRef] [PubMed]
18. Clarys, P.; Clijsen, R.; Taeymans, J.; Barel, A.O. Hydration measurements of the stratum corneum: Comparison between the capacitance method (digital version of the Corneometer CM 825®) and the impedance method (Skicon-200EX®). *Skin Res. Technol.* **2012**, *18*, 316–323. [CrossRef]
19. Barel, A.O.; Clarys, P. In vitro calibration of the capacitance method (Corneometer CM 825) and conductance method (Skicon-200) for the evaluation of the hydration state of the skin. *Skin Res. Technol.* **1997**, *3*, 107–113. [CrossRef]
20. Matias, A.R.; Ferreira, M.; Costa, P.; Neto, P. Skin colour, skin redness and melanin biometric measurements: Comparison study between Antera(®) 3D, Mexameter(®) and Colorimeter(®). *Skin Res. Technol.* **2015**, *21*, 346–362. [CrossRef]
21. Tupker, R.A.; Willis, C.; Berardksca, E.; Lee, C.H.; Fartasch, M.; Atinrat, T.; Serup, J. Guidelines on Sodium Lauryl Sulfate (SLS) Exposure Tests: A Report from the Standardization Group* of the European Society of Contact Dermatitis. *Contact Dermat.* **1997**, *37*, 53–69. [CrossRef]
22. Logger, J.G.M.; Olydam, J.I.; der Weg, W.W.; van Erp, P.E.J. Noninvasive Skin Barrier Assessment: Multiparametric Approach and Pilot Study. *Cosmetics* **2019**, *6*, 20. [CrossRef]
23. Qassem, M.; Kyriacou, P. Review of Modern Techniques for the Assessment of Skin Hydration. *Cosmetics* **2019**, *6*, 19. [CrossRef]
24. Green, M.; Feschuk, A.M.; Kashetsky, N.; Maibach, H.I. Normal TEWL-how it can be defined? A systematic review. *Exp. Dermatol.* **2022**, *31*, 1618–1631. [CrossRef]
25. Pilkington, S.M.; Bulfone-Paus, S.; Griffiths, C.E.M.; Watson, R.E.B. Inflammaging and the Skin. *J. Investig. Dermatol.* **2021**, *141*, 1087–1095. [CrossRef] [PubMed]
26. He, X.; Gao, X.; Xie, W. Research Progress in Skin Aging and Immunity. *Int. J. Mol. Sci.* **2024**, *25*, 4101. [CrossRef]
27. Pająk, J.; Nowicka, D.; Szepietowski, J.C. Inflammaging and Immunosenescence as Part of Skin Aging—A Narrative Review. *Int. J. Mol. Sci.* **2023**, *24*, 7784. [CrossRef]
28. Park, K.Y.; Ko, E.J.; Kim, I.S.; Li, K.; Kim, B.J.; Seo, S.J.; Kim, M.N.; Hong, C.K. The effect of evening primrose oil for the prevention of xerotic cheilitis in acne patients being treated with isotretinoin: A pilot study. *Ann. Dermatol.* **2014**, *26*, 706–712. [CrossRef]
29. Brosche, T.; Platt, D. Effect og borage oil consumption on fatty acid metabolism, transepidermal water loss and skin parameters in elderly people. *Arch. Gerontol. Geriatr.* **2000**, *30*, 139–150. [CrossRef]
30. Chiricozzi, A.; Belloni Fortina, A.; Galli, E. Current therapeutic paradigm in pediatric atopic dermatitis: Practical guidance from a national expert panel. *Allergol. Immunopathol.* **2019**, *47*, 194–206. [CrossRef]
31. Michalak, M.; Pierzak, M.; Kręcisz, B.; Suliga, E. Bioactive Compounds for Skin Health: A Review. *Nutrients* **2021**, *13*, 203. [CrossRef] [PubMed]
32. Michalak, M.; Kieltyka-Dadasiewicz, A. Oils from fruit seeds and their dietetic and cosmetic significance. *Herba Pol.* **2018**, *64*, 63–70. [CrossRef]
33. Burris, J.; Rietkerk, W.; Woolf, K. Acne: The role of medical nutrition therapy. *J. Acad. Nutr. Diet.* **2013**, *113*, 416–430. [CrossRef] [PubMed]
34. Marchlewicz, M.; Polakowska, Z.; Maciejewska-Markiewicz, D.; Stachowska, E.; Jakubiak, N.; Kiedrowicz, M.; Rak-Załuska, A.; Duchnik, M.; Wajs-Syrenicz, A.; Duchnik, E. Fatty Acid Profile of Erythrocyte Membranes in Patients with Psoriasis. *Nutrients* **2024**, *16*, 1799. [CrossRef] [PubMed]
35. Morita, A. Tobacco smoke causes premature skin aging. *J. Derm. Sci.* **2007**, *48*, 169–175. [CrossRef] [PubMed]
36. Lahmann, C.; Bergemann, J.; Harrison, G.; Young, A. Matrix metalloproteinase-1 and skin ageing in smokers. *Lancet* **2001**, *357*, 935–936. [CrossRef] [PubMed]
37. Alotaibi, G.F.; Alsalman, H.H.; Alhallaf, R.A.; Ahmad, R.A.; Alshareef, H.A.; Muammar, J.M.; Alsaif, F.M.; Alotaibi, F.F.; Balaha, M.F.; Ahmed, N.J.; et al. The Association of Smoking with Contact Dermatitis: A Cross-Sectional Study. *Healthcare* **2023**, *11*, 427. [CrossRef]
38. Hergesell, K.; Paraskevopoulou, A.; Opálka, L.; Velebný, V.; Vávrová, K.; Dolečková, I. The effect of long-term cigarette smoking on selected skin barrier proteins and lipids. *Sci. Rep.* **2023**, *13*, 11572. [CrossRef]

applied sciences

MDPI

Review

Intrinsic and Extrinsic Factors Associated with Hair Graying (Canities) and Therapeutic Potential of Plant Extracts and Phytochemicals

Yong Chool Boo [1,2,3]

[1] Department of Molecular Medicine, School of Medicine, Kyungpook National University, Daegu 41944, Republic of Korea; ycboo@knu.ac.kr
[2] BK21 Plus KNU Biomedical Convergence Program, Department of Biomedical Science, The Graduate School, Kyungpook National University, Daegu 41944, Republic of Korea
[3] Cell and Matrix Research Institute, Kyungpook National University, Daegu 41944, Republic of Korea

Abstract: This review aims to gain insight into the major causes of hair graying (canities) and how plant-derived extracts and phytochemicals could alleviate this symptom. Research articles on human hair graying were searched and selected using the PubMed, Web of Science, and Google Scholar databases. We first examined the intrinsic and extrinsic factors associated with hair graying, such as the reduced capacity of melanin synthesis and transfer, exhaustion of melanocyte stem cells (MSCs) and melanocytes, genetics and epigenetics, race, gender, family history, aging, oxidative stress, stress hormones, systematic disorders, nutrition, smoking, alcohol consumption, lifestyle, medications, and environmental factors. We also examined various plants and phytochemicals that have shown a potential to interfere with the onset or progression of human hair graying at different levels from in vitro studies to clinical studies: the extract of *Polygonum multiflorum* and its major components, 2,3,5,4′-tetrahydroxystilbene-2-*O*-β-D-glucoside and emodin; the extract of *Eriodictyon angustifolium* and its major flavonoid compounds, hydroxygenkwanin, sterubin, and luteolin; the extracts of Adzuki beans (*Vigna angularis*), Fuzhuan brick tea (*Camellia sinensis*), and *Gynostemma pentaphyllum*; bixin, a carotenoid compound found in *Bixa orellana*; and rhynchophylline, an alkaloid compound found in certain *Uncaria* species. Experimental evidence supports the notion that certain plant extracts and phytochemicals could alleviate hair graying by enhancing MSC maintenance or melanocyte function, reducing oxidative stress due to physiological and environmental influences, and managing the secretion and action of stress hormones to an appropriate level. It is suggested that hair graying may be reversible through the following tactical approaches: selective targeting of the p38 mitogen-activated protein kinase (MAPK)–microphthalmia-associated transcription factor (MITF) axis, nuclear factor erythroid 2-related factor 2 (NRF2), or the norepinephrine–β2 adrenergic receptor (β2AR)–protein kinase A (PKA) signaling pathway.

Keywords: hair graying; gray hair; canities; *Polygonum multiflorum*; *Eriodictyon angustifolium*; melanocyte stem cells (MSCs); microphthalmia-associated transcription factor (MITF); p38 mitogen-activated protein kinase (MAPK); oxidative stress; nuclear factor erythroid 2-related factor 2 (NRF2); norepinephrine; β2 adrenergic receptor (β2AR)

1. Introduction

Human hair does not continue to grow throughout life, but overlapping hair cycles occur approximately 25 times, resulting in changes in the overall hair population. The hair cycle consists of the anagen (growth), catagen (regression, intermediate, or transition), and telogen (resting) phases [1]. The final stage of the telogen phase is also separately classified as the exogen (shedding) phase [2]. The hair and associated cells and matrices constitute a mini-organ called a hair follicle [3,4]. Hair follicle keratinocyte stem cells (HFKSCs) of the keratinocyte lineage reside in the hair follicle bulge area [1,5,6]. The new hair germs at the telogen phase also contain HFKSCs or progenitor cells [7]. During the anagen onset, some primed HFKSCs (in the hair germ) and quiescent HFKSCs (in the bulge) are activated and migrate to the matrix, becoming transit-amplifying cells [5]. These cells continue dividing, proliferating, differentiating, and keratinizing, leading to hair production and growth [8].

Melanoblasts, derived from neural crest cells via melanoblasts/glial bipotent progenitor cells, differentiate into melanocytes. A portion of melanoblasts dedifferentiate to melanocyte stem cells (MSCs), which can differentiate into transit-amplifying cells and then into melanocytes [9]. MSCs found alongside HFKSCs in the hair follicle bulge are located in the secondary hair germs at the telogen phase [10,11]. During the anagen onset, some MSCs differentiate into mature melanocytes via transit-amplifying cells and migrate to the matrix area where melanocytes produce and supply melanin pigments to incorporate them into the hair fibers [10]. A recent study has reported that transit-amplifying cells may dedifferentiate into MSCs, while some MSCs may become directed into other pathways and may be unable to differentiate into melanocytes [11]. Overall, coordinated differentiation of stem cell and progenitor lineages in the hair follicle is important for hair growth and pigmentation [12]. Furthermore, hair growth and pigmentation are influenced by interactions with nerves, adipocytes, and immune cells neighboring hair follicles [13,14].

Hair graying, also known as canities, is the progressive loss of pigment in hair follicles, resulting in the appearance of gray or white hair [15]. It is a representative sign of human aging, along with hair loss [16]. Hair graying begins in the 40s for most people (senile hair graying) but can occur prematurely in some individuals (premature hair graying) [17,18]. As synchronous coordination of hair development and melanin synthesis is a requirement for normal hair pigmentation, its failure can cause hair graying [17]. Although the 50/50/50 rule of thumb, stating that "at age 50 years, 50% of the population has at least 50% of gray hair", is not necessarily valid or applicable to all races and genders, a significant number of people experience hair graying as they age [17,19–21]. Hair graying can occur even in young people when an insufficient amount of melanin is deposited within the growing hair shafts [22]. It causes psychological stress for some people that affects their mental health and social life [23].

Hair graying is a part of the aging process and is not generally considered a disease for most people unless closely related to a specific health condition or nutritional deficiency. In the past, hair graying was considered irreversible and had no proper countermeasure other than dyeing. However, the way we look at hair graying is changing as the understanding of the causative mechanism is deepening with new reports on cases of hair color restoration [24–27]. The repigmentation of white or gray hair is observed in patients receiving various medications [28,29]. A reversal of premature hair graying is also observed in patients treated with a topical formulation of 5% Melitane, an α-melanocyte-stimulating hormone (α-MSH) biomimetic peptide, or 2% Greyverse, an α-MSH biomimetic palmitoyl tetrapeptide-20, combined with oral supplements containing biotin and calcium pantothenate [30,31]. Therefore, we might be entering the era of medical therapy for hair graying [32,33].

This review aims to gain insight into the major causes of hair graying and how plant extracts and phytochemicals could alleviate the symptoms. Plants have unique survival strategies, distinct from animals, and synthesize secondary metabolites called phytochemicals [34]. Plant extracts have been used in traditional medicine, and some phytochemicals have been developed into active pharmaceutical ingredients or lead compounds for new drug development [35]. There is growing interest in the benefits and usability of plant-derived natural products in the cosmetics and dermatological fields [36–39]. We have recently reviewed the plant extracts and phytochemicals used to treat hair loss [40]. In line with this previous review, we explore the effects of plant extracts and phytochemicals on hair graying in the present review. In the context of existing review papers on the causes and treatments of hair graying [32,33], the present review focuses on deriving specific therapeutic molecular targets by interrelating studies on the causal mechanisms of hair graying and the biological activities of plant extracts and phytochemicals.

We first examine the latest studies on intrinsic and extrinsic factors associated with human hair graying. Then, we examine experimental studies demonstrating the therapeutic potential of plant extracts and phytochemicals in alleviating hair graying. Finally, we discuss major issues in human hair graying summarized by the following questions:

- What can be the targets for treating hair graying?
- What are the modulatory targets of plant extracts and phytochemicals?
- What should be the next research topics to advance treatments for human hair graying?

We hope this review will help identify novel therapeutic strategies and tactics for alleviating human hair graying.

2. Methods

The PubMed (https://pubmed.ncbi.nlm.nih.gov/), Web of Science (https://www.webofscience.com/), and Google Scholar (https://scholar.google.com/) databases were accessed on 4 July 2024 to search for research articles related to the topic of this narrative review. A preliminary search using several keywords, such as hair, grey, gray, graying, greying, white, canities, extract, plant, herb, phytochemical, etc., and Boolean search commands, such as 'AND', 'OR', and 'NOT', resulted in hundreds of articles. The search results were refined by limiting the search ranges for some keywords to title words only, followed by a brief reading of their titles and abstracts to select approximately 130 articles highly focused on human hair graying research. Most of these articles were cited and used for in-depth discussion in this review. The articles related to intrinsic and extrinsic factors associated with hair graying are mentioned mostly in Section 3. The articles related to the therapeutic potential of plant extracts and phytochemicals are mentioned mostly in Section 4.

3. Intrinsic and Extrinsic Factors Associated with Hair Graying

In this chapter, we examine various intrinsic and extrinsic factors directly or indirectly related to the onset and progression of hair graying. Intrinsic factors refer to biological factors, such as cellular, genetic, and physiological factors, whereas extrinsic factors refer to life factors comprising personal habits and external influences, as summarized in Figure 1. Since these factors are interconnected and influence each other, the classification provided here is arbitrary.

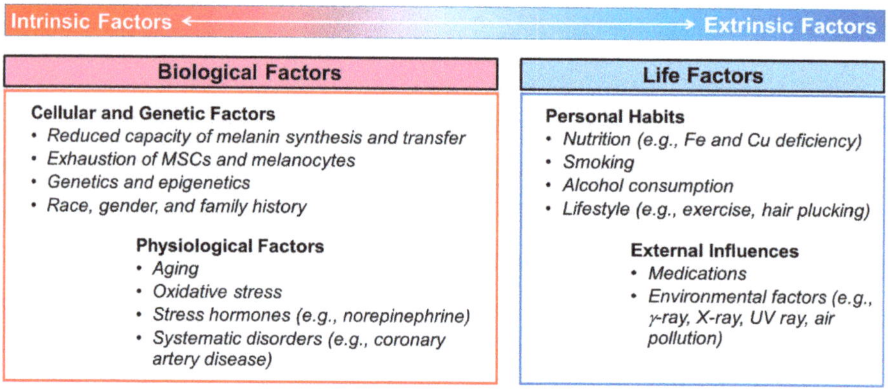

Figure 1. Intrinsic and extrinsic factors associated with the incidence or progression of hair graying in humans. Abbreviations: Cu, copper; Fe, iron; MSCs, melanocyte stem cells; UV, ultraviolet.

3.1. Reduced Capacity of Melanin Synthesis and Transfer

GeneChip Array analysis of mRNA expression profiles between pigmented, gray, and white scalp hair follicles from human donors identified 194 and 192 downregulated genes and 186 and 177 upregulated genes in gray and white hair follicles, respectively, compared to the pigmented hair follicles. The downregulated genes included tyrosinase (TYR), tyrosinase-related protein 1 (TYRP1), premelanosome protein (PMEL, Pmel17, GP100), melan-A (MLANA), KIT (c-Kit, receptor for stem cell factor), and MET (c-Met, receptor for hepatocyte growth factor) [41].

Genome-wide comparison of the gene expression profiles of gray and black hair follicular cells of human patients with premature hair graying identified 127 differentially expressed genes (DEGs). Many melanogenesis-associated proteins, such as TYR, MLANA, PMEL, TYRP1, solute carrier family 45 member 2 (SLC45A2), KIT, G protein-coupled receptor 143 (GPR143), and oculocutaneous albinism 2 (OCA2) were found to be downregulated in gray hair follicles compared to black hair follicles [42].

These studies support the view that the expression of genes related to melanin synthesis, melanosome biogenesis, and other melanocyte-relevant functions is reduced in hair follicles producing white or gray hair, resulting in insufficient melanin supply and reduced melanin incorporation into hair filaments during the anagen. Most DEGs in the above-mentioned studies are target genes of the transcriptional regulation by microphthalmia-associated transcription factor (MITF); hence, reduced MITF activity in melanocytes causes hair graying.

Placental sphingolipid induced the expression of MITF and TYR via the p38 mitogen-activated protein kinase (MAPK) signaling pathway in murine melanoma B16F10 cells and its topical application increased MITF expression in follicular melanocytes and the growth of fresh dense black hair in age-onset gray-haired C57BL/6J mice [43]. Therefore, the melanin synthesis and transport capacity of melanocytes, which are controlled by MITF, are the key variables most directly related to hair pigmentation and graying.

3.2. Exhaustion of MSCs and Melanocytes

Hair pigmentation can be affected by defects or depletion of MSCs, immature precursor cells, and mature melanocytes residing in the hair follicle bulge or bulb [44–46]. Loss of MSCs within the bulge niche preceded the loss of differentiated melanocytes in the hair matrix [47,48]. B-cell lymphoma 2 (BCL-2) deficiency caused selective apoptosis of MSCs, but not differentiated melanocytes, while MITF mutation accelerated pigmentation, differentiation, and senescence of MSCs [47]. Expressions of melanocyte markers, such as MITF-M, SRY-box transcription factor 10 (SOX10), paired box 3 (PAX3), pro-opiomelanocortin (POMC), TYR, TYRP1, and tyrosinase-related protein 2 (TYRP2), also called dopachrome tautomerase (DCT), were absent or reduced in the bulbs of white (non-pigmented) hair compared to black (pigmented) hair [49,50]. White hair shafts grew significantly faster than black hair shafts; the expression of keratins (KRTs) such as KRT6, KRT14, KRT16, KRT25, and KRT83 and keratin-associated protein (KRTAP) 4 isoforms was upregulated in white hair follicles compared to black hair follicles [51]. Therefore, the rapid growth of hair shafts may exhaust MSCs; hence, the incomplete self-maintenance of MSCs may be the causative mechanism for hair graying [47,52,53].

Various cell signaling pathways are involved in maintaining melanoblasts and MSCs in hair follicles [54]. Hair graying was observed in mice deficient in Notch1, Notch2 receptor, or RBP-J kappa transcription factor due to the disappearance of melanocytes in the hair matrix [55]. Therefore, Notch signaling might play an essential role in the melanocyte population during hair follicle cycles [56]. Phosphatase and tensin homolog (PTEN) encodes a phosphatidylinositol-3,4,5-trisphosphate 3-phosphatase which lowers phosphatidylinositol-3,4,5-trisphosphate levels in cells, thereby negatively regulating the phosphatidylinositol 3-kinase (PI3K)–protein kinase B (PKB) signaling pathway [57]. PTEN-deficient mice were resistant to MSC exhaustion in the hair follicle bulge and hair graying induced by repeated depilation in mice [58]. The increased expression of KIT ligand (KITL), also called stem cell factor (SCF), as well as of hepatocyte growth factor (HGF), and endothelin 3 (ET3), which promote melanocyte proliferation of melanoblasts and MSCs, alleviated hair graying induced by repeated hair plucking and shaving in C57BL/6 mice [59]. On the other hand, the expression of β-catenin in the anagen and telogen skin of aged mice was higher than that in young mice, and overexpression of Wingless and Int-1 (WNT) 10b promoted an excessive differentiation and depletion of MSCs, increasing hair graying [60]. Thus, the excessive consumption, ectopic differentiation, and aging of MSCs hinder the development of mature melanocytes in harmony with the hair follicle cycle, resulting in hair graying [61].

3.3. Genetics and Epigenetics

Although most people experience hair graying as they age, some people show symptoms of premature hair graying at a much earlier age [62,63]. The age criteria for classifying premature hair graying vary according to race [18,22,62]. Premature hair graying occurs mainly due to genetic factors, although it is also influenced by various other factors [22,64,65]. A familial history of premature hair graying is an important risk factor for the expression of this phenotype [66–68].

MITF plays a critical role at multiple stages of the melanocyte life cycle across species, with its expressional depletion resulting in a fully depigmented hair or coat [69]. The heterozygosity for MITF exacerbates hair graying in transgenic (*DctSox10*) mice that conditionally overexpress SOX10 by upregulating a type I interferon (IFN)-regulated gene expression signature associated with innate immunity [70].

Genome-wide association studies revealed that interferon regulatory factor 4 (IRF4) downregulation is associated with scalp hair graying [71]. IRF4 interacts with MITF and transcription factor AP-2α (TFAP2A), inducing gene expression of melanogenic enzymes, such as TYR [72]. Accordingly, hair-graying prediction models have been proposed by combining age, sex, and 13 unique single nucleotide polymorphisms (SNPs), including those in IRF4, maestro heat-like repeat family member 2A (MROH2A), kinesin family member 1A (KIF1A), and non-structural maintenance of chromosomes element 1 homolog (NSMCE1) [73].

A recent study has reported that the S519N variant of SAM and SH3 domain-containing protein 1 (SASH1) causes dysfunction in MSCs maintenance, contributing to the development of certain aging-related pigmentation disorders, namely senile lentigo of the skin and scalp hair graying in a Hispanic family [74]. SASH1 is a scaffold protein binding to tankyrase 2 (TNKS2) and regulating the division of MSCs, but its S519N variant does not function normally due to a low binding affinity to TNKS2 [74,75].

MicroRNAs, produced by the dicer (endoribonuclease type 3), are involved in lowering target gene expression by inducing mRNA degradation or suppressing translation [76]. Mimicking stress conditions, the inactivation of dicer causes the misplacement of melanocytes within the hair follicle, resulting in reduced melanin transfer to keratinocytes in the growing hair [77]. These effects were partly mediated by miR-92b, which targets integrin alpha chain 5 (ITGA5) involved in the regulation of melanocyte migration, implicating that such epigenetic mechanisms might link stress to premature hair graying.

From a mechanistic perspective, it is considered that certain genetic or epigenetic variations may cause acquired hair graying by negatively affecting MSC maintenance and melanocyte migration, whereas some genetic or epigenetic variations resulting in differences in the melanogenic function of melanocytes may affect natural hair color.

3.4. Race, Gender, and Family History

The age at which premature hair graying is defined varies by race: 20 in Caucasians, 25 in Asians, and 30 in African Americans [18,22,62]. This is because the genetic background and natural hair color of different races vary, and the time at which the results of hair graying become apparent varies [73]. Although natural hair color and brightness vary by race, there is no evidence that graying occurs more frequently or intensely in certain races [19].

In many races, men tend to develop gray hair earlier and more extensively than women of similar age [19]. Graying patterns also vary by gender [19,78]. In men, hair graying tends to occur first and is more prominent in the temporal region (the sides of the head). In women, the frontal and parietal regions (top and crown) are affected more often and early. In a survey study conducted on 620 gray-haired Korean subjects, 33.9% of men and 55.2% of women responded that they dye their hair regularly [79]. This means that women are relatively more concerned about hair graying.

Several studies have supported that hair graying is significantly affected by a hereditary component [66,67,73]. A cross-sectional survey of 100 young men less than 25 years old with premature hair graying showed that 39% of individuals had a family history, with 26% from the paternal side, 10% from the maternal side, and 3% from both parents [67]. It is noteworthy that a man's hair graying may be more influenced by his father's history than his mother's.

3.5. Aging

Aging is the major factor altering the expression of hair graying-associated genes [80]. Aged transgenic mice carrying the oncogenic *RET* gene exhibit gray hair due to an accelerated hair cycle, accompanied by the senescence of HFKSCs expressing ET1 and a decrease in follicular MSCs expressing endothelin receptor B (ETRB) [81]. Comparison of human scalp samples from young people aged 27–40 years with less gray hair and senior people aged 61–90 years with more gray hair confirmed the age-dependent increase in senescent follicular HFKSCs, decrease in ET-1 expression, and reduction in the number of functional follicular MSCs [81]. Thus, HFKSC senescence, MSC depletion, and their altered gene expression are considered the cause of age-dependent hair graying.

Comparison of gene expression between hair follicles of young and elder volunteers, aged below 25 and above 50 years, respectively, indicated that KRT33, KRT34, KRTAP4-4, and KRTAP4-7 became downregulated with age [82]. Lipidomic analysis of follicular tissues of black and white hair obtained from female participants aged 55–65 years indicated that white hair contains reduced total lipid, sphingolipid (particularly glucosylceramide and galactosylceramide), phosphatidylserine, and phosphatidic acid levels [83]. These studies suggest that age-dependent hair graying is accompanied or influenced by structural, biochemical, and pathophysiological changes in the hair interior and surrounding microenvironment.

3.6. Oxidative Stress

As people age, there is an increase in oxidative damage and a decrease in antioxidant defenses in the body [84]. Oxidative stress also plays a major role in age-induced hair graying via multiple effects on the viability and functionality of MSCs and mature melanocytes [85–87]. Cross-sectional studies support that systemic redox imbalance or oxidative stress can affect the incidence and degree of premature hair graying [88,89]. Specifically, serum malondialdehyde (MDA) levels increased while serum glutathione (GSH) and superoxide dismutase (SOD) levels decreased in patients with premature canities compared to controls [89].

Various factors, including cellular melanin synthesis reaction and ultraviolet (UV) irradiation, can cause the production of reactive oxygen species (ROS) in melanocytes [90,91]. Furthermore, the age-related decrease in BCL-2 expression makes melanocytes more vulnerable to oxidative stress and apoptotic death [92].

The reduction in catalase (CAT) expression and activity in hair follicles causes hydrogen peroxide (H_2O_2) accumulation, impairment of methionine sulfoxide repair, melanocyte dysfunction, and cell death [93,94]. In addition to CAT downregulation, hydroxyl radical scavenging activity becomes repressed in unpigmented hair follicles compared to normally pigmented hair follicles [50]. Ataxia telangiectasia mutated (ATM) protein, a serine/threonine protein kinase involved in the protection against oxidative DNA damage, is highly expressed within the nuclei of melanocytes in anagen hair follicle bulbs, but its expression is reduced proportionally to the degree of canities [95]. Thus, the balance between oxidative stress and antioxidant defense can affect human hair follicle growth, differentiation, and pigmentation.

The role of oxidative stress as the main causal mechanism in hair graying is also supported by hair depigmentation in animal models irradiated with UV, γ, or X-rays [96,97]. Hair graying can be induced by irradiating animals or organ-cultured hair follicles with γ-rays or X-rays, providing useful experimental models for hair graying research [97–99]. HFKSCs are more prone to oxidative DNA damage induced by X-ray irradiation than follicular MSCs [100]. When PUVA phototherapy (a combination treatment using psoralen and UVA) was applied to the backs of black mice after their hair had been removed, gray or white hair grew instead of black hair on the affected bare skin, and this effect was attenuated by pretreatment with a gel containing SOD [96]. Dietary restriction delayed hair graying in C57BL/6 J mice induced by γ-ray irradiation, but the effect was temporary [101]. Dietary restriction prolongs the quiescence of DNA-damaged hair follicle stem cells and does not prevent their eventual death and depletion.

3.7. Stress Hormones

The autonomic nervous system (ANS) and hypothalamic–pituitary–adrenal (HPA) axis, which are connected both anatomically and functionally, mediate acute and prolonged responses to stress signals, respectively [102]. The activation of sympathetic ANS and HPA triggers the release of catecholamines (e.g., adrenaline and norepinephrine) and glucocorticoid stress hormone (e.g., cortisol in humans, corticosterone in rodents), respectively, into the bloodstream [103].

Zhang et al. demonstrated that pain induced by the injection of resiniferatoxin, a capsaicin analog, increased serum norepinephrine and corticosterone levels and number of unpigmented white hairs in C57BL/6J mice with black coat color, and this change was inhibited by buprenorphine (an opioid analgesic) [104]. MSCs in the hair follicles express the glucocorticoid receptor (GR), the melanocortin 1 receptor (MC1R), and the β2 adrenergic receptor (β2AR), and the selective depletion of β2AR (but not GR depletion, MC1R depletion, or adrenalectomy) prevented white hair development, suggesting that the hyperactivation of sympathetic nerves causes the release of norepinephrine from the skin [104,105]. Additionally, the secretion of norepinephrine led to the rapid differentiation and eventual depletion of MSCs [106].

3.8. Systematic Disorders

Premature hair graying is associated with various disorders [18], such as coronary artery disease [107–109], obesity [66], metabolic diseases [110], hearing impairment [111], osteopenia [112], atopy [113], and autoimmune diseases including vitiligo [70]. Premature hair graying is a strong predictive marker of coronary artery disease, especially in smokers compared to non-smokers [114,115].

3.9. Nutrition

A study analyzing serum levels of trace elements in patients with premature hair graying demonstrated lower copper levels and higher iron levels in the patient group compared to the control group while zinc levels did not differ between the two groups [64]. In another study, iron, copper, and calcium were all lower in the patient group than in the control group [116]. An 11-year-old male with slowly graying hair had low serum ferritin and hemoglobin levels, and five months of iron supplementation (oral ferrous sulfate, 40 mg daily) normalized serum ferritin and hemoglobin levels and recovered black hair color in this patient [117]. Epidemiological or cross-sectional studies also suggested that premature hair graying is related to low serum ferritin and calcium levels [118–120]. On the other hand, serum levels of selenium and lead did not significantly differ between the two groups with and without premature hair graying [121].

Hair loss and graying hair were observed in patients receiving intravenous fat-free fluid for a long period. These symptoms were related to a deficiency of essential fatty acids and were cured by topical safflower oil containing high linoleic acid levels [122]. Several studies investigated the link between hair graying and certain vitamins, including vitamin B12 and vitamin D, and most, but not all, results supported a possible correlation [115,118–120,123,124].

3.10. Smoking, Alcohol Consumption, Lifestyle, Medications, and Environmental Factors

Smoking and alcohol consumption, which cause oxidative stress, are also independent risk factors correlated with hair graying [66,113,125–127]. A sedentary lifestyle, which can exacerbate metabolic diseases, affects hair pigmentation, whereas adequate sun exposure and regular exercise can help promote general health and prevent hair graying [22,65,123].

Hair plucking can occur when people shampoo, comb, dye, or perm your hair, and frequent repetition can deplete melanocytes and cause hair graying. Accordingly, repeated hair plucking is a technique to induce hair graying in animal models [11,59].

Various medications including certain chemotherapy drugs and antimalarial medications might cause hair graying [18]. Hair loss and graying were observed in patients during the treatment of depressive disorder with mirtazapine which fully recovered 10 weeks after medication discontinuation [128]. On the other hand, some anti-inflammatory drugs rarely promote the repigmentation of gray hair [28,29], but these cannot be applied for general treatment.

Hazardous environmental factors, such as X-rays, γ-rays, UV rays, and air pollution, can cause oxidative damage to hair follicle cells and hair filaments [129–131]. PUVA phototherapy or irradiation with X- or γ-rays can induce oxidative damage to cells and hair graying in animals, providing useful experimental models [96,97,101].

4. Therapeutic Potential of Plant Extracts and Phytochemicals

In this chapter, we review studies on the efficacy and mechanism of action of plant extracts and phytochemicals in various experimental hair-graying models.

4.1. Extracts and Bioactive Components of *Polygonum multiflorum*

Polygonum multiflorum (PM), also named *Pleuropterus multiflorus*, *Fallopia multiflora*, or *Reynoutria multiflora*, is the most widely used plant in the treatment of hair graying. This plant has been used in traditional medicine to treat various disorders [132]. It has two different medicinal forms, *Polygonum multiflorum* Radix (PMR) and *Polygonum multiflorum* Radix Preparata (PMRP), a processed form prepared by steaming PMR with black soybean decoction [133,134].

The main phytochemical components of PM can be classified into stilbenes, quinones, flavonoids, and others, as described in comprehensive reviews of this plant [135,136]. In high-performance liquid chromatography (HPLC) analysis, 2,3,5,4′-tetrahydroxystilbene-2-*O*-β-D-glucoside (TSG), emodin, emodin-8-*O*-β-D-glucopyranoside, physcion, and torachrysone-8-*O*-β-D-glucoside appeared as the major peaks [137,138]. Of these, TSG has been most extensively studied as the main active compound responsible for the pharmacological activities of PM [139,140].

Table 1 shows the biological activities of PM extracts and their phytochemical components evaluated in various experimental models to study melanin synthesis or hair pigmentation.

Table 1. Effects of the extracts and bioactive components of *Polygonum multiflorum* (PM) on melanin synthesis or hair pigmentation.

Models	Plant Extracts or Phytochemicals	Treatments	Efficacy	Molecular Targets	Reference
Six-week-old C57BL/6J male mice (10 animals per group); back furs bleached with 0.0375% H_2O_2 solution	PMR extract, PMRP extract, and TSG	TSG (oral 0.034 g kg^{-1} and topical 0.068 g kg^{-1}), PMR extract (oral 0.576 g kg^{-1} and topical 1.152 g kg^{-1}), or PMRP extract (oral 0.576 g kg^{-1} and topical 1.152 g kg^{-1}) for 6 weeks.	PMR extract, PMRP, and TSG reversed the total melanin content of hairs bleached by H_2O_2.	PMR extract, PMRP extract, and TSG reversed the expression levels of α-MSH, MC1R, and TYR downregulated by H_2O_2.	[134]
Primary human foreskin melanocytes cultured in vitro; human hair follicles cultured ex vivo.	Water extracts of PMRP and PMR.	Cells were treated with PMRP extract at 0.05, 0.1, 0.2, or 0.5 μg mL^{-1} in combination with 250 μM H_2O_2. Cells were treated with PMR extract at 0.01 mg mL^{-1} in combination with 350–1750 μM H_2O_2. Human hair follicles were treated with PMR extract at 0.1 mg mL^{-1} for 4 days.	PMR extract prevented cell death induced by H_2O_2. PMR extract enhanced pigmentation in human hair follicles ex vivo.	The water extract of PMRP reduced the intracellular ROS levels while preserving the GSH level in cells exposed to H_2O_2.	[141]
Human melanoma SKMEL-28 cell line; Zebrafish embryos and larvae	The methanolic extract of PM roots	SKMEL-28 cells were treated with the extract at 312, 625, and 2500 μg mL^{-1}. Zebrafish embryos were treated with the extract at 135 or 225 μg mL^{-1} for 4 days.	The extract increased the melanin contents in SKMEL-28 cells and zebrafish embryos and larvae.	The extract increased the mRNA levels of MC1R, MITF, and TYR in SKMEL-28 cells and zebrafish embryos and larvae.	[142]
Murine melanoma B16F10 cell line	The 70% ethanolic extract of PM roots	B16F10 cells were treated with the extract at 1, 2.5, 5, 10, 20, or 40 μM.	The extract increased the activity of TYR and the cellular melanin content without affecting cell viability.	The extract induced the expression of COX2, MITF, TYR, TYRP1, and TYRP2, and the phosphorylation of p38 MAPK in cells. Inhibition of p38 MAPK using SB203580 abolished the stimulatory effects of the extract on the expression of COX2, MITF, and TYR and melanin synthesis in cells.	[143]
Murine melanoma B16 cell line	TSG	B16 cells were treated with the extract at 1, 2, 5, or 10 μg mL^{-1} for 48 h.	TSG increased the activity of TYR and the cellular melanin content without affecting cell viability.	TSG induced the expression of TYR and MITF and the phosphorylation of CREB and p38 MAPK. Inhibition of p38 MAPK using SB203580 abolished the stimulatory effects of TSG on the expression of MITF and TYR and melanin synthesis in cells.	[144]
Murine melanoma B16F1 cell line exposed to H_2O_2	Emodin	B16F1 cells were treated with emodin at 0.125 0.25, 0.5, 1, or 2 μM in combination with 250 μM H_2O_2.	Emodin increased the activity of TYR and the melanin content in cells exposed to H_2O_2.	Emodin reduced the intracellular ROS levels and increased the protein levels of TYR, TYRP1, TYRP2, MITF, SIRT1, and FOXO1 while reducing those of ERK and phospho-ERK in cells treated with H_2O_2.	[145]
A randomized, double-blind clinical trial on 44 female subjects	APHG-1001: *Pueraria lobate, Pleuropterus multiflorus* (*Polygonum multiflorum*), and *Ginkgo biloba* leaves	Topical treatments with a tonic spray APHG-1001 twice daily for 24 weeks.	The tonic reduced the average number of newly developed gray hairs per unit area (cm^2) compared to the placebo group (6.3 versus 11.4) although without significant changes in the gross hair grayness.		[146]

Abbreviations: COX2, cyclooxygenase 2; CREB, cyclic AMP response element-binding protein; ERK, extracellular signal-regulated kinase; FOXO1, forkhead box protein O1; GSH, glutathione; H_2O_2, hydrogen peroxide; MAPK, mitogen-activated protein kinase; MC1R, melanocortin 1 receptor; MITF, microphthalmia-associated transcription factor; PM, *Polygonum multiflorum*; PMR, *Polygonum multiflorum* Radix; PMRP, *Polygonum multiflorum* Radix Preparata; ROS, reactive oxygen species; SIRT1, NAD-dependent deacetylase sirtuin-1; TSG, 2,3,5,4′-tetrahydroxy stilbene-2-*O*-β-D-glucoside; TYR, tyrosinase; TYRP1, tyrosinase-related protein 1; TYRP2, tyrosinase-related protein 2; α-MSH, α-melanocyte-stimulating hormone.

The hair-blackening effects of PMR extract, PMRP extract, and TSG in pure form were examined in 6-week-old C57BL/6J male mice (10 animals per group) with H_2O_2-induced hair bleaching [134]. For the model group, the 0.0375% H_2O_2 solution was spread on their back furs every morning for 6 weeks. Six hours after each H_2O_2 treatment, mice fasted for 2 h and received PMR extract (0.576 g kg^{-1} by gavage and 1.152 g kg^{-1} topically), PMRP extract (0.576 g kg^{-1} by gavage and 1.152 g kg^{-1} topically), or TSG (0.034 g kg^{-1} by gavage and 0.068 g kg^{-1} topically). The total melanin content of hairs was lowered by H_2O_2, which was reversed by PMR and PMRP extracts. When the expression of POMC, α-MSH, MC1R, agouti signaling protein (ASIP), MITF, TYR, TYRP1, and TYRP2 in the skin was analyzed by enzyme-linked immunosorbent assay (ELISA), only α-MSH, MC1R, and TYR levels were significantly downregulated by H_2O_2 and reversed by the PMR extract, PMRP extract, or TSG in the order of activity.

Sextius et al. reported the effects of PMR and PMRP extracts in protecting melanocytes from oxidative stress [141]. The water extract of PMRP reduced the intracellular ROS levels at 0.05, 0.1, 0.2, or 0.5 μg mL^{-1} while preserving the intracellular level of GSH at 0.05 μg mL^{-1} in primary human foreskin melanocytes exposed to 250 μM H_2O_2. PMR extract (0.01 mg mL^{-1}) prevented cell death induced by 1100–1750 μM H_2O_2. PMR extract (0.1 mg mL^{-1}) also enhanced pigmentation in human hair follicles ex vivo.

The methanolic extract of PM roots increased melanin contents and mRNA levels of MC1R, MITF, and TYR in human melanoma SKMEL-28 cells and the embryos and larvae of zebrafish at concentrations that did not cause cell death or morphological defects [142]. Considering that MC1R, MITF, and TYR mRNA levels were higher in the black hair follicles of young people than in their graying hair follicles, plant extracts that can upregulate the expression of MC1R, MITF, and TYR would help in the treatment of early hair graying and other pigmentation loss-related diseases.

The 70% ethanoic extract of PM roots (5 or 10 μg mL^{-1}) increased the melanin content in B16F10 cells without influencing cell viability [143]. The extract (10 μg mL^{-1}) increased the expression of cyclooxygenase-2 (COX2) and melanogenic genes, such as MITF, TYR, TYRP1, and TYRP2, at mRNA or protein levels. The extract induced p38 MAPK phosphorylation and SB203580 (inhibitor of p38 MAPK) abolished the expression of COX2, MITF, and TYR and melanin synthesis in cells stimulated by the extract of PM roots.

TSG (5 or 10 μg mL^{-1}) increased the TYR activity and melanin content in B16 cells without influencing cell viability or proliferation [144]. TSG (10 μg mL^{-1}) increased the mRNA and protein levels of TYR, which was associated with the phosphorylation (activation) of cyclic AMP response element-binding protein (CREB) and the increased protein levels of MITF. TSG induced p38 MAPK phosphorylation and SB203580 abolished the expression of MITF and TYR and melanin synthesis in TSG-stimulated cells.

Emodin reduced intracellular ROS levels and increased TYR activity and melanin synthesis in B16F1 cells exposed to 250 μM H_2O_2 [145]. It also increased TYR, TYRP1, TYRP2, MITF, NAD-dependent deacetylase sirtuin-1 (SIRT1), and forkhead box protein O1 (FOXO1) protein levels while decreasing those of Extracellular signal-regulated kinase (ERK) and phospho-ERK in H_2O_2-treated cells.

We could not find clinical trials evaluating the anti-hair-graying properties of PM extracts, but one clinical study investigated a hair tonic containing mixed extracts of several plants, including PM, *Pueraria lobate*, and *Ginkgo biloba* [146]. Although this tonic was not effective enough to change the gross hair grayness, it reduced the number of newly developed gray hairs per unit scalp area. Still, whether this effect is due to the phytochemicals of PM or other components remains unclear. Nonetheless, this study demonstrated the potential of herbal medicine to alleviate hair graying in humans.

4.2. Extracts and Bioactive Components of *Eriodictyon angustifolium*

Eriodictyon angustifolium (EA) is a plant called narrow-leaf Yerba Santa in Western America and has been used to treat respiratory and other diseases in traditional medicines and to mask the bitter flavor of pharmaceuticals [147]. It contains various flavonoid compounds (e.g., naringenin, eriodictyol, homoeriodictyol, hesperetin, sterubin, luteolin, diosmetin, chrysoeriol, hydroxygenkwanin, and isosakuranetin) and bisprenylated benzoic acid derivatives (e.g., erionic acids A, B, C, D, E, and F) [147–150].

In human gingival fibroblasts stimulated with LPS from *Porphyromonas gingivalis*, a crude ethanol/water extract of EA and its flavanone-rich fraction reduced the release of pro-inflammatory cytokines such as interleukin-6 (IL-6), IL-8, and macrophage chemoattractant protein-1 (MCP-1), supporting their anti-inflammatory potential [149]. Of individual flavanones, eriodictyol and naringenin had more potent anti-inflammatory effects than homoeriodictyol, sterubin, and hesperetin [149].

As summarized in Table 2, Taguchi et al. extensively studied the biological activities of EA extracts and their phytochemical components in cells, animal models, and human subjects, to examine their therapeutic potential in preventing hair graying.

Table 2. Effects of the extracts and bioactive components of *Eriodictyon angustifolium* (EA) on melanin synthesis or hair pigmentation.

Models	Plant Extracts or Phytochemicals	Treatments	Efficacy	Molecular Targets	Reference
Human melanoma HMVII cell line exposed to X-ray irradiation: NHEKs from a human subject with gray beard hair	Sterubin (a flavonoid compound found in EA)	Cells were treated with sterubin up to 100 μg mL^{-1}. Sterubin solution (0.1%) was topically applied to the beard area of a human subject for 4 weeks.	Sterubin increased melanin contents in HMVII cells and attenuated radiation-induced oxidative DNA damage (γH2AX foci) and death of NHEKs. Its topical application restored pigmentation in the gray beard hair of a human subject.	Sterubin increased the nuclear translocation of β-catenin, nuclear MITF, and TYR in HMVII cells. It also attenuated radiation-induced ROS production and mitochondrial membrane disintegration in NHEKs irradiated with X-ray.	[151]
Human melanoma HMVII cell line) exposed to X-ray irradiation: NHEKs from human subjects with gray beard hair (7 subjects per group) and human subjects with gray head hair (40–56 years old, 1 man and 9 women	Ethanol extracts of EA and EC	Cells were treated with EA or EC extract at 1–10 μg mL^{-1} (HMVII cell line) or 100 μg mL^{-1} (NHEKs). In clinical studies, the extracts (1 mL) were topically applied twice daily for one year (beard hair) or daily for 24 weeks (head hair).	The EA extract, but not the EC extract increased melanin contents in HMVII cells. The topical application of the EA extract reduced the occurrence of gray beard hair and head hair in human subjects.	EA extract, but not EC extract, increased the nuclear translocation of β-catenin, nuclear MITF, and TYR in melanocytes. Both EA and EC extracts attenuated radiation-induced DNA damage (γH2AX foci) and death of NHEKs.	[152]
C57BL/6 mice whose dorsal hairs were plucked and irradiated with X-ray	Hot water extract (tea) of EA leaves and stems	Three-week-old mice were given water or the extract freely for one month before hair plucking and X-ray irradiation.	The EA extract reduced the occurrence of gray hair one month later.	EA extract attenuated X-ray-induced oxidative DNA damage (γH2AX foci) of CD34+ HFKSCs in the skin.	[150]
NHEKs exposed to X-ray radiation: C57BL/6 mice subjected to X-ray irradiation or repeated hair plucking	Hydroxygenkwanin, sterubin, and luteolin (flavonoid compounds found in EA)	NHEKs were treated with each compound up to 100 μM. Mice were topically treated with each compound solution (0.1%) daily before or after X-ray irradiation or repeated hair plucking.	Hydroxygenkwanin pretreatment before X-ray irradiation or repeated hair plucking prevented hair graying in mice.	Hydroxygenkwanin, sterubin, and luteolin attenuated radiation-induced oxidative DNA damage (γH2AX foci) and death of NHEKs. Hydroxygenkwanin attenuated the radiation-induced ROS production and mitochondrial membrane disintegration.	[153]

Abbreviations: EA, *Eriodictyon angustifolium*; EC, *Eriodictyon californicum*; HFKSCs, hair follicular keratinocyte stem cells; NHEKs, normal human epidermal keratinocytes; γH2AX, H2A histone family member X phosphorylated at Ser139.

In one study [151], sterubin, one of the most abundant flavonoid components of EA [152], increased melanin in the human melanoma HMVII cell line, which was accompanied by an increase in β-catenin, MITF, and TYR. Sterubin also reduced the intracellular ROS level, oxidative DNA damage, mitochondrial damage, and death of normal human epidermal keratinocytes (NHEKs) irradiated with X-ray. Furthermore, the topical application of sterubin (0.1% solution) for 4 weeks restored pigmentation in the gray beard hair of a human subject.

In a subsequent study [152], the ethanol extract of EA increased melanin synthesis in the human melanoma HMVII cell line, which was accompanied by an increase in β-catenin, MITF, and TYR. However, the effects of *Eriodictyon californicum* (EC) extract were absent or weak. Additionally, EA extract reduced the oxidative DNA damage and death of NHEKs irradiated with X-ray more effectively than or similarly to EC extract. Furthermore, topical application of EA extract on the beard and scalp area for 1 year and 24 weeks, respectively, reduced the occurrence of gray hairs in human subjects.

As an animal model was fed water or plant extract freely for 1 month, its hair was plucked, irradiated with X-rays, and observed after 1 month, showing a reduced degree of gray hair in the group receiving the extract. When the skin was incised immediately after X-ray irradiation followed by histological analysis, the extract reduced the oxidative DNA damage in HFKSCs [150]. Thus, the extract might preserve the function of HFKSCs as a niche for MSCs, alleviating melanocyte depletion and hair graying.

When comparing the same molar concentrations of various flavonoid compounds, such as hydroxygenkwanin, sterubin, luteolin, eriodictyol, homoeriodictyol, hesperetin, and diosmetin, only hydroxygenkwanin, sterubin, and luteolin reduced the oxidative DNA damage and death of NHEKs irradiated with X-ray. Among them, hydroxygenkwanin had the strongest effect [153]. Hydroxygenkwanin further reduced the intracellular ROS level and mitochondrial damage in X-ray-irradiated NHEKs. When the animal model was topically treated with each compound for 7 days, followed by plucking its hair, irradiating it with X-rays, and observing it after 38 days, the degree of gray hair and oxidative DNA damage was reduced in the group treated with hydroxygenkwanin, sterubin, or luteolin. However, other flavonoids, such as eriodictyol, homoeriodictyol, hesperetin, and diosmetin, as well as a reference drug minoxidil, had no such effects. Furthermore, topical application of hydroxygenkwanin, sterubin, or luteolin reduced the hair graying induced by repeated hair plucking [153]. However, when hydroxygenkwanin was administered after X-ray irradiation, its inhibitory effects on hair graying were absent. Additionally, its inhibitory effects on hair graying were not seen in *Kit*-mutant mice, such as $Kit^{W/+}$ and $Kit^{V620Atg1/+}$ mice [154].

4.3. Other Plant Extracts and Phytochemicals

Table 3 introduces several studies on other plant extracts and phytochemicals.

Table 3. Effects of some plant extracts and phytochemicals on melanin synthesis or hair pigmentation.

Models	Plant Extracts or Phytochemicals	Treatments	Efficacy	Molecular Targets	Reference
Murine melanoma B16–BL6 cell line: Five to six-week-old C3H/HeJ Jel mice	A partially purified fraction from the hot water extract of adzuki beans (*Vigna angularis*) eluted through the Diaion HP-20 column	B16–BL6 cells were treated with the extract at 1, 2, or 3 mg mL^{-1} for 72 h. Mice were given water containing 1.0% extract for 12 weeks.	The extract increased melanin content and TYR activity in B16–BL6 cells without affecting cell proliferation. Its oral administration increased the melanin contents of back and abdomen hair in mice.	The extract increased cyclic AMP content and PKA activity in B16–BL6 cells. It increased the expression levels of TYR and TYRP2 but not TYRP1.	[155]
Melan-A cell line; Seven-week-old male C57BL/6 mice treated with hydroquinone	n-Hexane fraction of the ethanolic extract of Fuzhuan brick tea	Melan-A cells were treated with the fraction at 3, 10, or 30 μg mL^{-1}. The fraction was topically administered to mice at 50 or 100 mg kg^{-1} in combination with 100 mg kg^{-1} hydroquinone.	The fraction increased melanin content and TYR activity in Melan-A cells. Its topical application attenuated hair graying in mice.	The fraction increased the expression levels of NRF2 and its target genes, such as *HO1*, *SOD1*, *CAT*, and *GPX1* as well as *MITF*, *TYR*, *TYRP1*, and *TYRP2* in Melan-A cells and C57BL/6 mice. It also stimulated the phosphorylation of p38 MAPK, JNK, and ERK in Melan-A cells.	[156]
Murine melanoma B16 cell line stimulated with norepinephrine	Extract of GP leaves	B16 cells were treated with the extract at 0.1–6.4 mg mL^{-1} in combination with 1 μM norepinephrine for 48 h.	The extract increased melanin content and TYR activity in B16 cells stimulated with norepinephrine.	The extract upregulated MITF, TYR, TYRP1, and TYRP2 mRNA levels in cells.	[157]

Table 3. *Cont.*

Models	Plant Extracts or Phytochemicals	Treatments	Efficacy	Molecular Targets	Reference
C57BL/6J *Nrf2*$^{+/+}$ and *Nrf2*$^{-/-}$ mice; depilation and PUVA phototherapy	Bixin (an apocarotenoid compound found in the seeds of *Bixa orellana*)	Mice were topically applied with 1% bixin in polyethylene glycol 400 before PUVA phototherapy.	PUVA phototherapy caused hair graying in both *Nrf2*$^{+/+}$ and *Nrf2*$^{-/-}$ mice. Topical bixin pre-treatment attenuated hair graying in *Nrf2*$^{+/+}$ mice but paradoxically enhanced it in *Nrf2*$^{-/-}$ mice.	Topical bixin induced a twofold increase in NRF2 and its target genes, such as *TRXR1*, *p62 (SQSTM1)*, *NQO1*, *HO1*, *GCLM*, and *OGG1* in *Nrf2*$^{+/+}$ mice. PUVA phototherapy caused NRF2 activation, which was attenuated by bixin pretreatment.	[158]
Human A2058 cell line and murine melanoma B16F10 cell line stimulated with norepinephrine; seven-week-old male C57BL/6 mice; depilation and exposure to resiniferatoxin-induced stress	Rhynchophylline (an alkaloid compound found in certain *Uncaria* species)	Cells were treated with rhynchophylline at 50 or 100 μM in combination with 100 μM norepinephrine. In animal experiments, 1 mg mL^{-1} rhynchophylline (200 μL) was topically applied to the skin of mice twice a day for 30 days.	Rhynchophylline restored the melanogenic pathway and intracellular calcium balance and attenuated apoptosis of A2058 and B16F10 cells influenced by norepinephrine. Its topical application reduced the hair graying in the stressed mice.	Rhynchophylline exhibited strong and stable binding within the active site of β2AR and inhibited the norepinephrine–β2AR–PKA signaling pathway.	[159]

Abbreviations; CAT, catalase; ERK, extracellular signal-regulated kinase; GCLM, glutamate-cysteine ligase's modifier subunit; GP, *Gynostemma pentaphyllum*; GPX1, glutathione peroxidase 1; HO1, heme oxygenase 1; JNK, c-jun N-terminal kinase; MAPK, mitogen-activated protein kinase; MITF, microphthalmia-associated transcription factor; NQO1, NAD(P)H quinone oxidoreductase 1; NRF2, nuclear factor erythroid 2-related factor 2; OGG1, 8-oxoguanine DNA glycosylase; PKA, protein kinase A; PUVA, psoralen plus UVA; SOD1, superoxide dismutase 1; SQSTM1, sequestosome 1; TRXR1, thioredoxin reductase 1; TYR, tyrosinase; TYRP1, tyrosinase-related protein 1; TYRP2, tyrosinase-related protein 1; β2AR, β2 adrenergic receptor.

Adzuki beans (*Vigna angularis*), also called red beans, have a higher polyphenol content than mung beans or soybeans [160]. The partially purified fraction from the hot water extract of Adzuki beans activated MITF through the Protein kinase A (PKA) signaling pathway at the cellular level, increasing the expression of melanogenic enzymes and melanin synthesis in B16 cells [155]. This effect was also reproduced in animal experiments, as the extract increased the melanin contents of back and abdomen hair in mice [155]

Tea (*Camellia sinensis*) is classified into six major categories, green tea, white tea, yellow tea, oolong tea, black tea, and dark tea, in the order of increases in the fermentation process, with Fuzhuan brick tea representing a kind of dark tea [161]. The n-hexane fraction of the ethanolic extract of Fuzhuan brick tea (phytochemical components: gallic acid, theaflavin, theobromine, caffeine, epicatechin, and quercetin) stimulated both the nuclear factor erythroid 2-related factor 2 (NRF2) signaling pathways and the MAPK signaling pathways, leading to expression of heme oxygenase 1 (HO1), SOD1, CAT, glutathione peroxidase 1 (GPX1), MITF, TYR, TYRP1, and TYRP2, as well as melanin synthesis in Melan-A cells [156]. It also stimulated hair pigmentation in seven-week-old male C57BL/6 mice treated with hydroquinone.

The extract of *Gynostemma pentaphyllum* (GP) may help prevent both hair loss and graying. It stimulated the proliferation of human dermal papilla cells by upregulating the WNT signaling pathway while downregulating the transforming growth factor-β (TGF-β) signaling pathway. Furthermore, its topical application enhanced hair growth in mice [157]. The extract stimulated the proliferation of murine melanoma B16 cells and their expression of MITF, MITF, TYR, TYRP1, and TYRP2 [157]. It further increased melanin content and TYR activity in B16 cells in the presence of epinephrine [157].

Bixin is an apocarotenoid compound found in *Bixa orellana* that can stimulate the NRF2 pathway by modifying cysteine residue in kelch-like ECH-associated protein 1 (KEAP1) [162]. Its systemic administration reduced the oxidative DNA damage and inflammatory reaction in wild-type SKH-1 mice, but such effects were not seen in *Nrf2*$^{(-/-)}$ mice [162]. Topical bixin treatment reduced UV-induced oxidative DNA damage induced by UV in wild-type SKH-1 mice, but not in *Nrf2*$^{(-/-)}$ mice [158]. Additionally, topical bixin retarded hair graying induced by PUVA phototherapy in wild-type C57BL/6J mice, but not in *Nrf2*$^{(-/-)}$ mice [158]. Thus, bixin can alleviate photo-oxidative skin damage or hair graying through NRF2 activation.

Rhynchophylline is an alkaloid compound found in certain *Uncaria* species (Rubiaceae) [163]. Rhynchophylline was identified as a blocker of β2AR, inhibiting the norepinephrine–β2AR–PKA signaling pathway [159]. Rhynchophylline restored the melanogenic pathway and intracellular calcium balance and attenuated the apoptosis of the A2058 and murine melanoma B16F10 cell lines influenced by norepinephrine. Its topical application reduced hair graying in seven-week-old male C57BL/6 mice exposed to resiniferatoxin [159].

5. Discussion

5.1. What Can Be the Targets for Treating Hair Graying?

Premature or senile hair graying occurs when hair growth is relatively normal, but pigmentation is insufficient. Therefore, the most direct cause of hair graying is the decline in the viability and function of melanocytes that synthesize and supply melanin [164].

MITF performs a central regulatory function in the differentiation and proliferation of melanocytes and melanin synthesis [69,165]. Therefore, various factors can influence hair graying if they are upstream regulators of MITF, downstream targets of MITF, or interacting partners of MITF. The SCF/c-KIT, ET1/ETRB, WNT/Frizzled, and α-MSH/MC1R signaling pathways are involved in MITF regulation [80]. Mutations and expression levels of MITF and several proteins that interact with it, such as SOX10 and IRF4, can affect the differentiation of MSCs and the death and function of melanocytes, leading to hair graying [70,71]. Topical α-MSH mimetic peptides have been proven effective in alleviating hair graying in patients [30,31]. Topical placental sphingolipid increased MITF level in follicular melanocytes and black hair growth in age-onset gray-haired C57BL/6J mice [43].

Oxidative stress is caused by the overproduction of ROS or other pro-oxidants and a lack of antioxidants or antioxidant enzymes, resulting in oxidative damage to the key cellular molecules, such as nucleic acids, lipids, and proteins [89,95]. Oxidative stress is both a cause and a consequence of aging and various diseases. Poor nutrition, lack of exercise, smoking, alcohol consumption, radiation, and air pollution can all increase oxidative stress, accelerating hair graying [85–89]. Oxidative stress due to X-ray irradiation, PUVA phototherapy, H_2O_2, or hydroquinone can induce hair graying, which symptoms can be alleviated by various antioxidant strategies: inhibition of ROS generation, scavenging ROS using antioxidants or antioxidant enzymes, and increasing the antioxidant capacity of cells [96,101,156,158].

Stress hormones are the main triggers of hair graying [104,105], while male hormones are one of the main triggers of alopecia [166]. An injection of resiniferatoxin causes pain stress and hair graying in animal models by stimulating norepinephrine secretion from the affected area [104,105]. The norepinephrine–β2AR–PKA signaling pathway mediates the pain stress responses in the body, including rapid differentiation and depletion of MSCs in the hair follicles, leading to hair graying [104,105]. A natural alkaloid compound, rhynchophylline, was identified as a novel β2AR blocker and its biological activity antagonizing the action of norepinephrine was verified in cells [159]. In animal models, rhynchophylline alleviated resiniferatoxin-induced hair graying [159]. Repeated hair plucking also caused hair graying in animal models probably via pain induction, and several EA-derived flavonoid compounds reduced the occurrence of white or gray hair [104,153].

Taken together, pharmacological approaches enhancing the expression and activity of MITF in melanocytes or effectively mitigating the negative effects of oxidative stress and stress hormones are expected to provide useful strategies for treating hair graying.

5.2. What Are the Modulatory Targets of Plant Extracts and Phytochemicals?

As summarized in Figure 2, various plant extracts and phytochemicals activate MITF, alleviate oxidative stress, or block stress hormone-induced response in melanocytes.

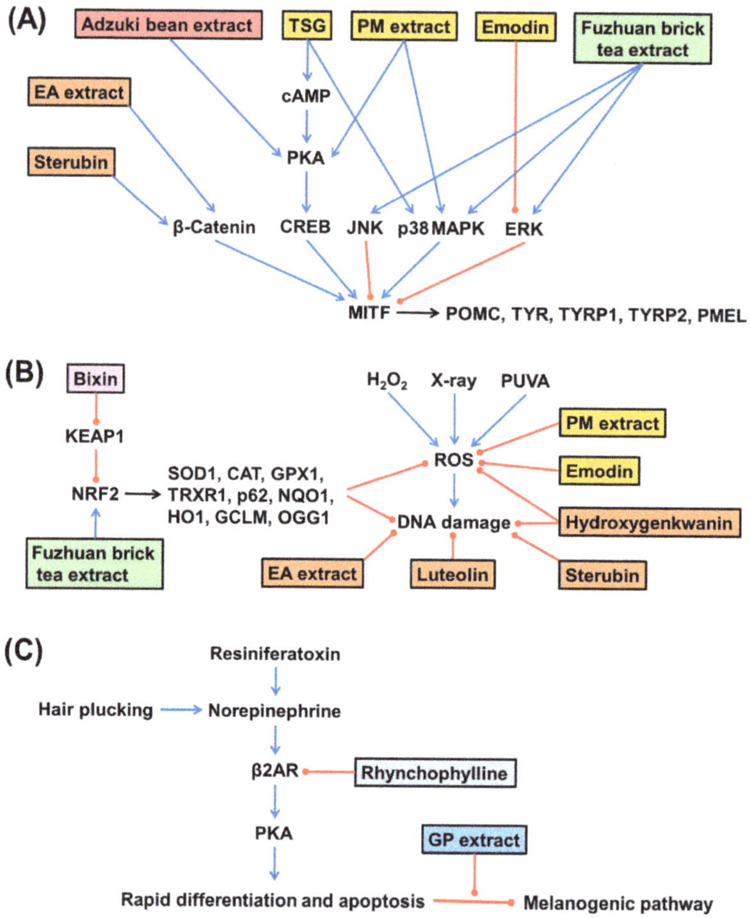

Figure 2. Effects of plant extracts and phytochemicals on the MITF-melanogenic pathways (**A**), ROS-mediated oxidative stress (**B**), and the stress hormone-mediated response (**C**) in melanocytic cells and animal models. Extracts and phytochemicals derived from the same plant are indicated with the same color boxes. The stimulatory and suppressive effects are indicated by blue sharp arrows and red blunt arrows, respectively. Abbreviations; CAT, catalase; CREB, cyclic AMP response element-binding protein; EA, *Eriodictyon angustifolium*; ERK, extracellular signal-regulated kinase; GCLM, glutamate-cysteine ligase's modifier subunit; GP, *Gynostemma pentaphyllum*; GPX1, glutathione peroxidase 1; HO1, heme oxygenase 1; JNK, c-jun N-terminal kinase; KEAP1, kelch-like ECH-associated protein 1; MAPK, mitogen-activated protein kinase; MITF, microphthalmia-associated transcription factor; NQO1, NAD(P)H quinone oxidoreductase 1; NRF2, nuclear factor erythroid 2-related factor 2; OGG1, 8-oxoguanine DNA glycosylase; PKA, protein kinase A; PM, *Polygonum multiflorum*; PMEL, premelanosome protein; POMC, pro-opiomelanocortin; PUVA, psoralen plus UVA; ROS, reactive oxygen species; SOD1, superoxide dismutase 1; TRXR1, thioredoxin reductase 1; TSG, 2,3,5,4′-tetrahydroxystilbene-2-*O*-β-D-glucoside; TYR, tyrosinase; TYRP1, tyrosinase-related protein 1; TYRP2, tyrosinase-related protein 1; β2AR, β2 adrenergic receptor.

The extract of PM and its phytochemical components, such as TSG and emodin, the extract of EA and its components, such as sterubin, and the extracts of Adzuki beans and Fuzhuan brick tea activated MITF in melanocytes. PM extract, TSG, and emodin stimulated p38 MAPK or PKA activation or suppressed ERK activation, increasing the expression of

MITF and target genes impacted by H_2O_2 [143–145]. EA extract and sterubin activated the β-catenin-dependent WNT signaling pathway and increased the expression of MITF and target genes impacted by X-ray irradiation [151,152]. Adzuki bean extract activated PKA and increased the expression of MITF and target genes [155]. The extract of Fuzhuan brick tea stimulated the phosphorylation of p38 MAPK, JNK, and ERK, which was associated with an increased expression of MITF and target genes [156].

The extracts of PM and its component (emodin), the extract of EA and its components (hydroxygenkwaninm, sterubin, and luteolin), the extract of Fuzhuan brick tea, and a carotenoid compound, bixin, all could alleviate oxidative stress in cells. The PM extract and emodin reduced the H_2O_2-induced cellular ROS levels [141,145]. The EA extract and its components, such as hydroxygenkwanin, sterubin, and luteolin, reduced the intracellular ROS level, oxidative DNA damage, and mitochondrial damage induced by X-ray irradiation [150,151,153]. The extract of Fuzhuan brick tea and bixin activated NRF2 and alleviated oxidative stress triggered by hydroquinone treatment or PUVA phototherapy [156,158].

Moreover, the extract of GP leaves and an alkaloid compound, rhynchophylline, blocked the stress hormone-induced response in cells. They inhibited the action of norepinephrine or resiniferatoxin and restored the melanogenesis pathway [157,159].

5.3. What Should Be the Next Research Topics to Advance Treatments for Human Hair Graying?

PKA and MAPKs play important roles in directly or indirectly regulating MITF expression and activity [167,168]. PKA activates CREB, which increases MITF expression in melanocytes stimulated by α-MSH [169]. Of the MAPKs, p38 MAPK directly phosphorylates MITF at Ser307 to increase its transcriptional activity [170] or activates PKA or other protein kinases, such as MAP kinase-activated protein kinase 2 (MAPKAPK2), to contribute to increased MITF expression through CREB [43,171]. ERK phosphorylates MITF at Ser73, enhancing both the transcriptional activity and proteasomal degradation of MITF, up- or downregulating MITF activity and expression depending on the cellular context [167]. JNK phosphorylates CREB-regulated transcription coactivator 3 (CRTC3), thereby inhibiting nuclear translocation and transcriptional activity, and suppressing MITF expression [172]. Thus, p38 MAPK has a consistent and greater influence than JNK and ERK in regulating MITF expression and activity.

The PM extract and TSG increased p38 MAPK activation and MITF expression, and this effect was neutralized by a selective inhibitor of p38 MAPK (SB203580). Additionally, placental sphingolipid and recombinant human IL-33 increased MITF expression through a similar mechanism [32,155]. These observations highlight the importance of p38 MAPK in regulating MITF expression and activity. Therefore, future studies should focus on the selective targeting of p38 MAPK as a strategy for treating hair graying due to melanocyte dysfunction.

NRF2 is a transcription factor regulating the gene expression of Phase II metabolism/antioxidant enzymes by binding the antioxidant-responsive elements (AREs) in the promoter of target genes, enhancing the cellular antioxidant defense capacity [173]. In basal state cells, cytosolic NRF2 is captured by dimeric KEAP1 or other proteins, which promote ubiquitylation and proteasomal degradation of NRF2. Under stimulated conditions, NRF2 escapes from KEAP1 or proteasomal degradation and translocates to the nucleus where it induces the transcription of target genes [174].

Topical administration of a fraction of Fuzhuan brick tea extract or an apocarotenoid compound bixin to mice increased the expression of NRF2 and its target genes in the skin and attenuated hair graying induced by hydroquinone treatment or PUVA phototherapy, respectively [156,158]. Therefore, an effective NRF2 activator could help attenuate hair graying due to oxidative stress. Consequently, future studies need to systemically examine the effects on hair graying of many known and unknown NRF2 modulators [174].

The rapid growth or turnover of hair might cause hair graying due to accelerated differentiation and depletion of MSCs [47,52,53]. Supporting this notion, less pigmented hair grew faster than pigmented black hair; in other words, hair that grew faster was less pigmented [51]. Although the cause of accelerated differentiation and depletion of MSCs during rapid hair growth is unclear, it might be due to excessive activation of the β-catenin-dependent WNT signaling pathway [60]. If so, the β-catenin-dependent WNT signaling pathway cannot be an ideal therapeutic target for alleviating hair graying, although it could be a promising therapeutic target for promoting hair growth [175,176].

Resiniferatoxin injection and repeated hair plucking commonly induce hair graying, probably by stimulating the release of norepinephrine, a stress hormone [104,153]. Interestingly, norepinephrine promoted hair follicle growth in an organotypic skin culture model [177], despite accelerating the differentiation and depletion of MSCs [106]. Therefore, proper regulation of the norepinephrine–β2AR–PKA signaling pathway is important for maintaining a balance between hair growth and pigmentation. Understanding how to control the secretion and action of norepinephrine to treat hair graying and hair loss will be an intriguing topic of future research.

Only a few studies have evaluated the anti-hair-graying efficacy of plant extracts, mixtures, or bioactive compounds through clinical trials [146,151,152], and most studies were conducted using human cells or mice. There are limitations to predicting their therapeutic efficacy for human hair graying based on mouse model studies because of the higher complexity of controlling hair pigmentation in humans. There may be other similar studies not covered in this review. For example, glycosylated polyphenols derived from *Larix europea* wood extract and *Camellia sinensis* leaf extract reduced oxidative stress and protected MSCs in human hair follicles ex vivo, and in a clinical study carried out for 4 months on 44 Caucasian male volunteers with 30% to 95% gray hair, a lotion containing the active ingredient reduced the proportion of gray hairs to the total hairs and the number of gray hairs per unit scalp area relative to day 0 values to a greater extent compared to the placebo control [178]. More expanded clinical trials are needed to evaluate the efficacy of various plant-derived natural products on human hair graying.

Early diagnosis and patient-tailored treatment are essential. As the possible causative factors for hair graying are known, specifying the exact cause through questionnaires and biochemical analysis is important for specific patients. In reported cases, hair graying occurring due to iron or essential fatty acid deficiency was treated by supplying each nutrient [117,122]. Mirtazapine-induced hair graying could be treated by discontinuing this medication [128]. Analysis of the patient's oxidative stress status and stress hormone levels will also provide an important guide to patient-tailored treatment.

Since plant-derived natural products have both pros and cons as medicines, they must be used for appropriate purposes in appropriate amounts to find a balance between efficacy and side effects [37,39,40]. For example, PM extract may be used to alleviate both hair loss and graying [133,134,179], but caution should be taken because its high doses can cause liver toxicity [180,181]. Molecular targets for treating hair graying are being specified, and information on the efficacy and mechanism of action of individual phytochemicals is expanding. We hope that by using such information, an effective treatment that will help maintain the health and beauty of hair will be developed in the near future.

6. Conclusions

This review addressed various intrinsic and extrinsic factors influencing the onset or progression of premature and senile hair graying. It further addressed the efficacy and mechanism of action of numerous plant extracts and phytochemicals that can help alleviate these symptoms. Certain types of hair graying can be prevented or treated by enhancing MSC maintenance or melanocyte function, reducing oxidative stress, and managing the secretion and action of stress hormones. Tactical approaches to pursue this goal may include a selective activation of the p38 MAPK–MITF axis, enhancing cellular antioxidant capacity through activating NRF2, and modulating the norepinephrine–β2AR–PKA signaling pathway. Furthermore, accurately diagnosing the cause of hair graying for specific patients and applying customized treatment as early as possible are important.

Funding: This research received no external funding.

Conflicts of Interest: The author declares no conflicts of interest.

References

1. Morgan, B.A. The Dermal Papilla: An Instructive Niche for Epithelial Stem and Progenitor Cells in Development and Regeneration of the Hair Follicle. *CSH Perspect. Med.* **2014**, *4*, a015180. [CrossRef]
2. Houschyar, K.S.; Borrelli, M.R.; Tapking, C.; Popp, D.; Puladi, B.; Ooms, M.; Chelliah, M.P.; Rein, S.; Pförringer, D.; Thor, D.; et al. Molecular Mechanisms of Hair Growth and Regeneration: Current Understanding and Novel Paradigms. *Dermatology* **2020**, *236*, 271–280. [CrossRef] [PubMed]
3. Lin, X.Y.; Zhu, L.; He, J. Morphogenesis, Growth Cycle and Molecular Regulation of Hair Follicles. *Front. Cell Dev. Biol.* **2022**, *10*, 899095. [CrossRef] [PubMed]
4. Schneider, M.R.; Schmidt-Ullrich, R.; Paus, R. The Hair Follicle as a Dynamic Miniorgan. *Curr. Biol.* **2009**, *19*, R132–R142. [CrossRef] [PubMed]
5. Zhang, B.; Chen, T. Local and systemic mechanisms that control the hair follicle stem cell niche. *Nat. Rev. Mol. Cell Biol.* **2024**, *25*, 87–100. [CrossRef]
6. Morita, R.; Sanzen, N.; Sasaki, H.; Hayashi, T.; Umeda, M.; Yoshimura, M.; Yamamoto, T.; Shibata, T.; Abe, T.; Kiyonari, H.; et al. Tracing the origin of hair follicle stem cells. *Nature* **2021**, *594*, 547. [CrossRef]
7. Lee, J.H.; Choi, S. Deciphering the molecular mechanisms of stem cell dynamics in hair follicle regeneration. *Exp. Mol. Med.* **2024**, *56*, 110–117. [CrossRef]
8. Natarelli, N.; Gahoonia, N.; Sivamani, R.K. Integrative and Mechanistic Approach to the Hair Growth Cycle and Hair Loss. *J. Clin. Med.* **2023**, *12*, 893. [CrossRef]
9. Mort, R.L.; Jackson, I.J.; Patton, E.E. The melanocyte lineage in development and disease. *Development* **2015**, *142*, 620–632. [CrossRef]
10. Huang, L.L.; Zuo, Y.Z.; Li, S.L.; Li, C.Y. Melanocyte stem cells in the skin: Origin, biological characteristics, homeostatic maintenance and therapeutic potential. *Clin. Transl. Med.* **2024**, *14*, e1720. [CrossRef]
11. Sun, Q.; Lee, W.; Hu, H.; Ogawa, T.; De Leon, S.; Katehis, I.; Lim, C.H.; Takeo, M.; Cammer, M.; Taketo, M.M.; et al. Dedifferentiation maintains melanocyte stem cells in a dynamic niche. *Nature* **2023**, *616*, 774. [CrossRef] [PubMed]
12. Wu, S.J.; Yu, Y.; Liu, C.Y.; Zhang, X.; Zhu, P.Y.; Peng, Y.; Yan, X.Y.; Li, Y.; Hua, P.; Li, Q.F.; et al. Single-cell transcriptomics reveals lineage trajectory of human scalp hair follicle and informs mechanisms of hair graying. *Cell Discov.* **2022**, *8*, 49. [CrossRef]
13. Chen, J.; Zheng, Y.X.; Hu, C.; Jin, X.X.; Chen, X.P.; Xiao, Y.; Wang, C.C. Hair Graying Regulators Beyond Hair Follicle. *Front. Physiol.* **2022**, *13*, 839859. [CrossRef] [PubMed]
14. Chen, J.; Fan, Z.X.; Zhu, D.C.; Guo, Y.L.; Ye, K.; Dai, D.; Guo, Z.; Hu, Z.Q.; Miao, Y.; Qu, Q. Emerging Role of Dermal White Adipose Tissue in Modulating Hair Follicle Development During Aging. *Front. Cell Dev. Biol.* **2021**, *9*, 728188. [CrossRef]
15. Singal, A.; Daulatabad, D.; Grover, C. Graying severity score: A useful tool for evaluation of premature canities. *Indian Dermatol. Online J.* **2016**, *7*, 164–167. [CrossRef]
16. Mirmirani, P. Age-related hair changes in men: Mechanisms and management of alopecia and graying. *Maturitas* **2015**, *80*, 58–62. [CrossRef]
17. O'Sullivan, J.D.B.; Nicu, C.; Picard, M.; Cheret, J.; Bedogni, B.; Tobin, D.J.; Paus, R. The biology of human hair greying. *Biol. Rev. Camb. Philos. Soc.* **2021**, *96*, 107–128. [CrossRef] [PubMed]
18. Kumar, A.B.; Shamim, H.; Nagaraju, U. Premature Graying of Hair: Review with Updates. *Int. J. Trichology* **2018**, *10*, 198–203. [CrossRef] [PubMed]
19. Panhard, S.; Lozano, I.; Loussouarn, G. Greying of the human hair: A worldwide survey, revisiting the '50' rule of thumb. *Br. J. Dermatol.* **2012**, *167*, 865–873. [CrossRef]
20. Acer, E.; Arslantas, D.; Emiral, G.Ö.; Ünsal, A.; Atalay, B.I.; Göktas, S. Clinical and epidemiological characteristics and associated factors of hair graying: A population-based, cross-sectional study in Turkey. *An. Bras. Dermatol.* **2020**, *95*, 439–446. [CrossRef]
21. Tobin, D.J.; Paus, R. Graying: Gerontobiology of the hair follicle pigmentary unit. *Exp. Gerontol.* **2001**, *36*, 29–54. [CrossRef] [PubMed]
22. Kaur, K.; Kaur, R.; Bala, I. Therapeutics of premature hair graying: A long journey ahead. *J. Cosmet. Dermatol.* **2019**, *18*, 1206–1214. [CrossRef] [PubMed]
23. Saad, M.; Babar, N.F.; Majeed, R.; Rehman, A.U.; Khan, O.A.; Chatha, D.E.; Aamir, U.; Nadeem, A.; Abbas, S. Impact of Premature Greying of Hair on Socio-cultural Adjustment and Self-esteem among Medical Undergraduates in Foundation University, Islamabad. *Cureus* **2019**, *11*, e5083. [CrossRef] [PubMed]
24. Rosenberg, A.M.; Rausser, S.; Ren, J.; Mosharov, E.V.; Sturm, G.; Ogden, R.T.; Patel, P.; Kumar Soni, R.; Lacefield, C.; Tobin, D.J.; et al. Quantitative mapping of human hair greying and reversal in relation to life stress. *Elife* **2021**, *10*, e67437. [CrossRef]
25. Pandhi, D.; Khanna, D. Premature graying of hair. *Indian J. Dermatol. Venereol. Leprol.* **2013**, *79*, 641–653. [CrossRef]
26. Chow, E.Y.; Salopek, T.G. Acitretin-Induced Repigmentation of Gray Hair: A Case Report. *Cureus* **2024**, *16*, e58261. [CrossRef]
27. Tian, Z.; Liu, Q.; Wang, Y.; Yang, D. Two case studies of persistent white hair regrowth after alopecia areata turning pigmented following treatment. *J. Cosmet. Dermatol.* **2024**, *23*, 2490–2495. [CrossRef]

28. Dasanu, C.A.; Mitsis, D.; Alexandrescu, D.T. Hair repigmentation associated with the use of lenalidomide: Graying may not be an irreversible process! *J. Oncol. Pharm. Pract.* **2013**, *19*, 165–169. [CrossRef]
29. Yale, K.; Juhasz, M.; Atanaskova Mesinkovska, N. Medication-Induced Repigmentation of Gray Hair: A Systematic Review. *Skin. Appendage Disord.* **2020**, *6*, 1–10. [CrossRef]
30. Sakhiya, J.J.; Sakhiya, D.J.; Patel, M.R.; Daruwala, F.R. Case report on premature hair graying treated with Melitane 5% and oral hair supplements. *Indian J. Pharmacol.* **2019**, *51*, 346–349. [CrossRef]
31. Chavan, D. Reversal of Premature Hair Graying Treated with a Topical Formulation Containing α-Melanocyte-Stimulating Hormone Agonist (Greyverse Solution 2%). *Int. J. Trichology* **2022**, *14*, 207–209. [CrossRef]
32. Feng, Z.R.; Qin, Y.; Jiang, G. Reversing Gray Hair: Inspiring the Development of New Therapies Through Research on Hair Pigmentation and Repigmentation Progress. *Int. J. Biol. Sci.* **2023**, *19*, 4588–4607. [CrossRef] [PubMed]
33. Paus, R.; Sevilla, A.; Grichnik, J.M. Human Hair Graying Revisited: Principles, Misconceptions, and Key Research Frontiers. *J. Investig. Dermatol.* **2024**, *144*, 474–491. [CrossRef] [PubMed]
34. Kaur, S.; Samota, M.K.; Choudhary, M.; Choudhary, M.; Pandey, A.K.; Sharma, A.; Thakur, J. How do plants defend themselves against pathogens—Biochemical mechanisms and genetic interventions. *Physiol. Mol. Biol. Plants* **2022**, *28*, 485–504. [CrossRef]
35. Nasim, N.; Sandeep, I.S.; Mohanty, S. Plant-derived natural products for drug discovery: Current approaches and prospects. *Nucleus* **2022**, *65*, 399–411. [CrossRef] [PubMed]
36. Boo, Y.C. Emerging Strategies to Protect the Skin from Ultraviolet Rays Using Plant-Derived Materials. *Antioxidants* **2020**, *9*, 637. [CrossRef]
37. Boo, Y.C. Can Plant Phenolic Compounds Protect the Skin from Airborne Particulate Matter? *Antioxidants* **2019**, *8*, 379. [CrossRef]
38. Sitarek, P.; Kowalczyk, T.; Wieezfinska, J.; Merecz-Sadowska, A.; Górski, K.; Sliwinski, T.; Skala, E. Plant Extracts as a Natural Source of Bioactive Compounds and Potential Remedy for the Treatment of Certain Skin Diseases. *Curr. Pharm. Des.* **2020**, *26*, 2859–2875. [CrossRef]
39. Boo, Y.C. Insights into How Plant-Derived Extracts and Compounds Can Help in the Prevention and Treatment of Keloid Disease: Established and Emerging Therapeutic Targets. *Int. J. Mol. Sci.* **2024**, *25*, 1235. [CrossRef]
40. Choi, J.Y.; Boo, M.Y.; Boo, Y.C. Can Plant Extracts Help Prevent Hair Loss or Promote Hair Growth? A Review Comparing Their Therapeutic Efficacies, Phytochemical Components, and Modulatory Targets. *Molecules* **2024**, *29*, 29102288. [CrossRef]
41. Peters, E.M.; Liezmann, C.; Spatz, K.; Ungethum, U.; Kuban, R.J.; Daniltchenko, M.; Kruse, J.; Imfeld, D.; Klapp, B.F.; Campiche, R. Profiling mRNA of the graying human hair follicle constitutes a promising state-of-the-art tool to assess its aging: An exemplary report. *J. Investig. Dermatol.* **2013**, *133*, 1150–1160. [CrossRef]
42. Bian, Y.M.; Wei, G.; Song, X.; Yuan, L.; Chen, H.Y.; Ni, T.; Lu, D.R. Global downregulation of pigmentation-associated genes in human premature hair graying. *Exp. Ther. Med.* **2019**, *18*, 1155–1163. [CrossRef] [PubMed]
43. Saha, B.; Singh, S.K.; Mallick, S.; Bera, R.; Datta, P.K.; Mandal, M.; Roy, S.; Bhadra, R. Sphingolipid-mediated restoration of Mitf expression and repigmentation in vivo in a mouse model of hair graying. *Pigment. Cell Melanoma Res.* **2009**, *22*, 205–218. [CrossRef] [PubMed]
44. Commo, S.; Gaillard, O.; Bernard, B.A. Human hair greying is linked to a specific depletion of hair follicle melanocytes affecting both the bulb and the outer root sheath. *Br. J. Dermatol.* **2004**, *150*, 435–443. [CrossRef] [PubMed]
45. Peters, E.M.; Imfeld, D.; Graub, R. Graying of the human hair follicle. *J. Cosmet. Sci.* **2011**, *62*, 121–125.
46. Pss, R.; Madhunapantula, S.V.; Betkerur, J.B.; Bovilla, V.R.; Shastry, V. Melanogenesis Markers Expression in Premature Graying of Hair: A Cross-Sectional Study. *Skin Pharmacol. Physiol.* **2022**, *35*, 180–186. [CrossRef]
47. Nishimura, E.K.; Granter, S.R.; Fisher, D.E. Mechanisms of hair graying: Incomplete melanocyte stem cell maintenance in the niche. *Science* **2005**, *307*, 720–724. [CrossRef]
48. Ji, J.; Ho, B.S.; Qian, G.; Xie, X.M.; Bigliardi, P.L.; Bigliardi-Qi, M. Aging in hair follicle stem cells and niche microenvironment. *J. Dermatol.* **2017**, *44*, 1097–1104. [CrossRef]
49. Choi, Y.J.; Yoon, T.J.; Lee, Y.H. Changing expression of the genes related to human hair graying. *Eur. J. Dermatol.* **2008**, *18*, 397–399. [CrossRef]
50. Shi, Y.; Luo, L.F.; Liu, X.M.; Zhou, Q.; Xu, S.Z.; Lei, T.C. Premature Graying as a Consequence of Compromised Antioxidant Activity in Hair Bulb Melanocytes and Their Precursors. *PLoS ONE* **2014**, *9*, e93589. [CrossRef]
51. Choi, H.I.; Choi, G.I.; Kim, E.K.; Choi, Y.J.; Sohn, K.C.; Lee, Y.; Kim, C.D.; Yoon, T.J.; Sohn, H.J.; Han, S.H.; et al. Hair greying is associated with active hair growth. *Br. J. Dermatol.* **2011**, *165*, 1183–1189. [CrossRef] [PubMed]
52. Steingrímsson, E.; Copeland, N.G.; Jenkins, N.A. Melanocyte stem cell maintenance and hair graying. *Cell* **2005**, *121*, 9–12. [CrossRef]
53. Jo, S.K.; Lee, J.Y.; Lee, Y.; Kim, C.D.; Lee, J.H.; Lee, Y.H. Three Streams for the Mechanism of Hair Graying. *Ann. Dermatol.* **2018**, *30*, 397–401. [CrossRef] [PubMed]
54. Wang, X.X.; Liu, Y.H.; He, J.; Wang, J.R.; Chen, X.D.; Yang, R.H. Regulation of signaling pathways in hair follicle stem cells. *Burns Trauma* **2022**, *10*, tkac022. [CrossRef] [PubMed]
55. Schouwey, K.; Delmas, V.; Larue, L.; Zimber-Strobl, U.; Strobl, L.J.; Radtke, F.; Beermann, F. Notch1 and Notch2 receptors influence progressive hair graying in a dose-dependent manner. *Dev. Dyn.* **2007**, *236*, 282–289. [CrossRef] [PubMed]
56. Schouwey, K.; Beermann, F. The Notch pathway: Hair graying and pigment cell homeostasis. *Histol. Histopathol.* **2008**, *23*, 609–619. [CrossRef]

57. Lee, Y.R.; Pandolfi, P.P. PTEN Mouse Models of Cancer Initiation and Progression. *Cold Spring Harb. Perspect. Med.* **2020**, *10*, a037283. [CrossRef]
58. Inoue-Narita, T.; Hamada, K.; Sasaki, T.; Hatakeyama, S.; Fujita, S.; Kawahara, K.; Sasaki, M.; Kishimoto, H.; Eguchi, S.; Kojima, I.; et al. Pten deficiency in melanocytes results in resistance to hair graying and susceptibility to carcinogen-induced melanomagenesis. *Cancer Res.* **2008**, *68*, 5760–5768. [CrossRef]
59. Endou, M.; Aoki, H.; Kobayashi, T.; Kunisada, T. Prevention of hair graying by factors that promote the growth and differentiation of melanocytes. *J. Dermatol.* **2014**, *41*, 716–723. [CrossRef]
60. Zhang, Z.; Lei, M.; Xin, H.; Hu, C.; Yang, T.; Xing, Y.; Li, Y.; Guo, H.; Lian, X.; Deng, F. Wnt/β-catenin signaling promotes aging-associated hair graying in mice. *Oncotarget* **2017**, *8*, 69316–69327. [CrossRef]
61. Zhang, X.; Zhu, J.; Zhang, J.; Zhao, H. Melanocyte stem cells and hair graying. *J. Cosmet. Dermatol.* **2023**, *22*, 1720–1723. [CrossRef]
62. Phan, B.; Ali, A.M.; Black, T.A.; Kashyap, A.; Niazi, M.; Rashid, R.M. Fractal Pattern in the Premature Graying of Hair: A Case Report. *Cureus* **2024**, *16*, e59994. [CrossRef] [PubMed]
63. Rangu, S.A.; Oza, V.S. Poliosis, hair pigment dilution, and premature graying of the hair: A diagnostic approach in pediatric patients and review of the literature. *Pediatr. Dermatol.* **2024**, *41*, 197–203. [CrossRef]
64. Fatemi Naieni, F.; Ebrahimi, B.; Vakilian, H.R.; Shahmoradi, Z. Serum iron, zinc, and copper concentration in premature graying of hair. *Biol. Trace Elem. Res.* **2012**, *146*, 30–34. [CrossRef] [PubMed]
65. Nath, B.; Gupta, V.; Kumari, R. A Community Based Study to Estimate Prevalence and Determine Correlates of Premature Graying of Hair among Young Adults in Srinagar, Uttarakhand, India. *Int. J. Trichology* **2020**, *12*, 206–212. [CrossRef] [PubMed]
66. Shin, H.; Ryu, H.H.; Yoon, J.; Jo, S.; Jang, S.; Choi, M.; Kwon, O.; Jo, S.J. Association of premature hair graying with family history, smoking, and obesity: A cross-sectional study. *J. Am. Acad. Dermatol.* **2015**, *72*, 321–327. [CrossRef] [PubMed]
67. Anggraini, D.R.; Feriyawati, L.; Hidayat, H.; Wahyuni, A.S. Risk Factors Associated with Premature Hair Greying of Young Adult. *Open Access Maced. J. Med. Sci.* **2019**, *7*, 3762–3764. [CrossRef]
68. Thompson, K.G.; Marchitto, M.C.; Ly, B.C.K.; Chien, A.L. Evaluation of Physiological, Psychological, and Lifestyle Factors Associated with Premature Hair Graying. *Int. J. Trichology* **2019**, *11*, 153–158. [CrossRef]
69. Steingrímsson, E.; Arnheiter, H.; Hallsson, J.H.; Lamoreux, M.L.; Copeland, N.G.; Jenkins, N.A. Interallelic complementation at the mouse Mitf locus. *Genetics* **2003**, *163*, 267–276. [CrossRef]
70. Harris, M.L.; Fufa, T.D.; Palmer, J.W.; Joshi, S.S.; Larson, D.M.; Incao, A.; Gildea, D.E.; Trivedi, N.S.; Lee, A.N.; Day, C.P.; et al. A direct link between MITF, innate immunity, and hair graying. *PLoS Biol.* **2018**, *16*, e2003648. [CrossRef]
71. Adhikari, K.; Fontanil, T.; Cal, S.; Mendoza-Revilla, J.; Fuentes-Guajardo, M.; Chacón-Duque, J.C.; Al-Saadi, F.; Johansson, J.A.; Quinto-Sanchez, M.; Acuña-Alonzo, V.; et al. A genome-wide association scan in admixed Latin Americans identifies loci influencing facial and scalp hair features. *Nat. Commun.* **2016**, *7*, 10815. [CrossRef] [PubMed]
72. Praetorius, C.; Grill, C.; Stacey, S.N.; Metcalf, A.M.; Gorkin, D.U.; Robinson, K.C.; Van Otterloo, E.; Kim, R.S.Q.; Bergsteinsdottir, K.; Ogmundsdottir, M.H.; et al. A Polymorphism in IRF4 Affects Human Pigmentation through a Tyrosinase-Dependent MITF/TFAP2A Pathway. *Cell* **2013**, *155*, 1022–1033. [CrossRef] [PubMed]
73. Pospiech, E.; Kukla-Bartoszek, M.; Karlowska-Pik, J.; Zielinski, P.; Wozniak, A.; Boron, M.; Dabrowski, M.; Zubanska, M.; Jarosz, A.; Grzybowski, T.; et al. Exploring the possibility of predicting human head hair greying from DNA using whole-exome and targeted NGS data. *BMC Genom.* **2020**, *21*, 538. [CrossRef] [PubMed]
74. Lambert, K.A.; Clements, C.M.; Mukherjee, N.; Pacheco, T.R.; Shellman, S.X.; Henen, M.A.; Vogeli, B.; Goldstein, N.B.; Birlea, S.; Hintzsche, J.; et al. SASH1 S519N Variant Links Skin Hyperpigmentation and Premature Hair Graying to Dysfunction of Melanocyte Lineage. *J. Investig. Dermatol.* **2024**, S0022-0202X(0024)00393-00392. [CrossRef]
75. Lambert, K.A.; Clements, C.M.; Mukherjee, N.; Pacheco, T.R.; Shellman, S.X.; Henen, M.A.; Vogeli, B.; Goldstein, N.B.; Birlea, S.; Hintzsche, J.; et al. SASH1 interacts with TNKS2 and promotes human melanocyte stem cell maintenance. *bioRxiv* **2023**. [CrossRef]
76. Yao, Q.; Chen, Y.; Zhou, X. The roles of microRNAs in epigenetic regulation. *Curr. Opin. Chem. Biol.* **2019**, *51*, 11–17. [CrossRef]
77. Bertrand, J.U.; Petit, V.; Aktary, Z.; de la Grange, P.; Elkoshi, N.; Sohier, P.; Delmas, V.; Levy, C.; Larue, L. Loss of Dicer in Newborn Melanocytes Leads to Premature Hair Graying and Changes in Integrin Expression. *J. Investig. Dermatol.* **2024**, *144*, 601–611. [CrossRef]
78. Jo, S.J.; Paik, S.H.; Choi, J.W.; Lee, J.H.; Cho, S.; Kim, K.H.; Eun, H.C.; Kwon, O.S. Hair Graying Pattern Depends on Gender, Onset Age and Smoking Habits. *Acta Derm.-Venereol.* **2012**, *92*, 160–161. [CrossRef]
79. Jo, S.J.; Shin, H.; Paik, S.H.; Choi, J.W.; Lee, J.H.; Cho, S.; Kwon, O. The Pattern of Hair Dyeing in Koreans with Gray Hair. *Ann. Dermatol.* **2013**, *25*, 401–404. [CrossRef]
80. Wang, S.; Kang, Y.; Qi, F.; Jin, H. Genetics of hair graying with age. *Ageing Res. Rev.* **2023**, *89*, 101977. [CrossRef]
81. Iida, M.; Tazaki, A.; Yajima, I.; Ohgami, N.; Taguchi, N.; Goto, Y.; Kumasaka, M.Y.; Prevost-Blondel, A.; Kono, M.; Akiyama, M.; et al. Hair graying with aging in mice carrying oncogenic RET. *Aging Cell* **2020**, *19*, e13273. [CrossRef] [PubMed]
82. Giesen, M.; Gruedl, S.; Holtkoetter, O.; Fuhrmann, G.; Koerner, A.; Petersohn, D. Ageing processes influence keratin and KAP expression in human hair follicles. *Exp. Dermatol.* **2011**, *20*, 759–761. [CrossRef] [PubMed]
83. Ma, Y.; He, C. Exploration of potential lipid biomarkers for age-induced hair graying by lipidomic analyses of hair shaft roots with follicular tissue attached. *J. Cosmet. Dermatol.* **2022**, *21*, 6118–6123. [CrossRef]

84. Tan, B.L.; Norhaizan, M.E.; Liew, W.P.; Sulaiman Rahman, H. Antioxidant and Oxidative Stress: A Mutual Interplay in Age-Related Diseases. *Front. Pharmacol.* **2018**, *9*, 1162. [CrossRef]
85. Seiberg, M. Age-induced hair greying—The multiple effects of oxidative stress. *Int. J. Cosmet. Sci.* **2013**, *35*, 532–538. [CrossRef]
86. Trueb, R.M. Oxidative stress in ageing of hair. *Int. J. Trichology* **2009**, *1*, 6–14. [CrossRef]
87. Arck, P.C.; Overall, R.; Spatz, K.; Liezman, C.; Handjiski, B.; Klapp, B.F.; Birch-Machin, M.A.; Peters, E.M.J. Towards a "free radical theory of graying": Melanocyte apoptosis in the aging human hair follicle is an indicator of oxidative stress induced tissue damage. *FASEB J.* **2006**, *20*, 1567–1569. [CrossRef]
88. Acer, E.; Kaya Erdogan, H.; Kocaturk, E.; Saracoglu, Z.N.; Alatas, O.; Bilgin, M. Evaluation of oxidative stress and psychoemotional status in premature hair graying. *J. Cosmet. Dermatol.* **2020**, *19*, 3403–3407. [CrossRef] [PubMed]
89. Saxena, S.; Gautam, R.K.; Gupta, A.; Chitkara, A. Evaluation of Systemic Oxidative Stress in Patients with Premature Canities and Correlation of Severity of Hair Graying with the Degree of Redox Imbalance. *Int. J. Trichology* **2020**, *12*, 16–23. [CrossRef]
90. Chen, J.J.; Liu, Y.; Zhao, Z.; Qiu, J. Oxidative stress in the skin: Impact and related protection. *Int. J. Cosmet. Sci.* **2021**, *43*, 495–509. [CrossRef]
91. Kaminski, K.; Kazimierczak, U.; Kolenda, T. Oxidative stress in melanogenesis and melanoma development. *Wspolczesna Onkol./Contemp. Oncol.* **2022**, *26*, 112447. [CrossRef]
92. Geueke, A.; Mantellato, G.; Kuester, F.; Schettina, P.; Nelles, M.; Seeger, J.M.; Kashkar, H.; Niemann, C. The anti-apoptotic Bcl-2 protein regulates hair follicle stem cell function. *EMBO Rep.* **2021**, *22*, e52301. [CrossRef] [PubMed]
93. Wood, J.M.; Decker, H.; Hartmann, H.; Chavan, B.; Rokos, H.; Spencer, J.D.; Hasse, S.; Thornton, M.J.; Shalbaf, M.; Paus, R.; et al. Senile hair graying: H_2O_2-mediated oxidative stress affects human hair color by blunting methionine sulfoxide repair. *FASEB J.* **2009**, *23*, 2065–2075. [CrossRef] [PubMed]
94. Kauser, S.; Westgate, G.E.; Green, M.R.; Tobin, D.J. Human Hair Follicle and Epidermal Melanocytes Exhibit Striking Differences in Their Aging Profile which Involves Catalase. *J. Investig. Dermatol.* **2011**, *131*, 979–982. [CrossRef]
95. Sikkink, S.K.; Mine, S.; Freis, O.; Danoux, L.; Tobin, D.J. Stress-sensing in the human greying hair follicle: Ataxia Telangiectasia Mutated (ATM) depletion in hair bulb melanocytes in canities-prone scalp. *Sci. Rep.* **2020**, *10*, 18711. [CrossRef]
96. Emerit, I.; Filipe, P.; Freitas, J.; Vassy, J. Protective effect of superoxide dismutase against hair graying in a mouse model. *Photochem. Photobiol.* **2004**, *80*, 579–582. [CrossRef] [PubMed]
97. Ungvari, A.; Kiss, T.; Gulej, R.; Tarantini, S.; Csik, B.; Yabluchanskiy, A.; Mukli, P.; Csiszar, A.; Harris, M.L.; Ungvari, Z. Irradiation-induced hair graying in mice: An experimental model to evaluate the effectiveness of interventions targeting oxidative stress, DNA damage prevention, and cellular senescence. *Geroscience* **2024**, *46*, 3105–3122. [CrossRef]
98. Dai, D.M.; He, Y.; Guan, Q.; Fan, Z.X.; Zhu, Y.M.; Wang, J.; Wu, S.L.; Chen, J.; Le, D.J.; Hu, Z.Q.; et al. Modeling human gray hair by irradiation as a valuable tool to study aspects of tissue aging. *Geroscience* **2023**, *45*, 1215–1230. [CrossRef]
99. Anderson, Z.T.; Palmer, J.W.; Idris, M.I.; Villavicencio, K.M.; Le, G.; Cowart, J.; Weinstein, D.E.; Harris, M.L. Topical RT1640 treatment effectively reverses gray hair and stem cell loss in a mouse model of radiation-induced canities. *Pigment Cell Melanoma Res.* **2021**, *34*, 89–100. [CrossRef]
100. Aoki, H.; Hara, A.; Motohashi, T.; Kunisada, T. Keratinocyte stem cells but not melanocyte stem cells are the primary target for radiation-induced hair graying. *J. Investig. Dermatol.* **2013**, *133*, 2143–2151. [CrossRef]
101. Qiu, R.; Qiu, X.; Su, M.; Sun, M.; Wang, Y.; Wu, J.; Wang, H.; Tang, D.; Tao, S. Dietary Restriction Delays But Cannot Heal Irradiation-Induced Hair Graying by Preserving Hair Follicle Stem Cells in Quiescence. *Rejuvenation Res.* **2023**, *26*, 242–252. [CrossRef]
102. Mueller, B.; Figueroa, A.; Robinson-Papp, J. Structural and functional connections between the autonomic nervous system, hypothalamic-pituitary-adrenal axis, and the immune system: A context and time dependent stress response network. *Neurol. Sci.* **2022**, *43*, 951–960. [CrossRef] [PubMed]
103. Ulrich-Lai, Y.M.; Herman, J.P. Neural regulation of endocrine and autonomic stress responses. *Nat. Rev. Neurosci.* **2009**, *10*, 397–409. [CrossRef] [PubMed]
104. Zhang, B.; Ma, S.; Rachmin, I.; He, M.; Baral, P.; Choi, S.; Goncalves, W.A.; Shwartz, Y.; Fast, E.M.; Su, Y.; et al. Hyperactivation of sympathetic nerves drives depletion of melanocyte stem cells. *Nature* **2020**, *577*, 676–681. [CrossRef] [PubMed]
105. Zhang, B.; He, M.G.; Rachmin, I.; Yu, X.L.; Kim, S.; Fisher, D.E.; Hsu, Y.C. Melanocortin 1 receptor is dispensable for acute stress induced hair graying in mice. *Exp. Dermatol.* **2021**, *30*, 572–577. [CrossRef] [PubMed]
106. Mendelsohn, A.R.; Larrick, J.W. The Danger of Being Too Sympathetic: Norepinephrine in Alzheimer's Disease and Graying of Hair. *Rejuvenation Res.* **2020**, *23*, 68–72. [CrossRef]
107. Kocaman, S.A.; Cetin, M.; Durakoglugil, M.E.; Erdogan, T.; Canga, A.; Cicek, Y.; Dogan, S.; Sahin, I.; Satiroglu, O.; Bostan, M. The degree of premature hair graying as an independent risk marker for coronary artery disease: A predictor of biological age rather than chronological age. *Anadolu Kardiyol. Derg.* **2012**, *12*, 457–463. [CrossRef]
108. Sharma, K.; Humane, D.; Shah, K.; Patil, S.; Charaniya, R.; Meniya, J. Androgenic alopecia, premature graying, and hair thinning as independent predictors of coronary artery disease in young Asian males. *Cardiovasc. Endocrinol.* **2017**, *6*, 152–158. [CrossRef]
109. ElFaramawy, A.A.A.; Hanna, I.S.; Darweesh, R.M.; Ismail, A.S.; Kandil, H.I. The degree of hair graying as an independent risk marker for coronary artery disease, a CT coronary angiography study. *Egypt. Heart J.* **2018**, *70*, 15–19. [CrossRef]
110. Paik, S.H.; Jang, S.; Joh, H.K.; Lim, C.S.; Cho, B.; Kwon, O.; Jo, S.J. Association Between Premature Hair Greying and Metabolic Risk Factors: A Cross-sectional Study. *Acta Derm.-Venereol.* **2018**, *98*, 748–752. [CrossRef]

111. Ozbay, I.; Kahraman, C.; Kucur, C.; Namdar, N.D.; Oghan, F. Is there a relationship between premature hair greying and hearing impairment? *J. Laryngol. Otol.* **2015**, *129*, 1097–1100. [CrossRef] [PubMed]
112. Agarwal, S.; Choudhary, A.; Kumar, A.; Zaidi, A.; Mohanty, S.; Yadav, S. A Study of Association of Premature Graying of Hair and Osteopenia in North Indian Population. *Int. J. Trichology* **2020**, *12*, 75–78. [CrossRef] [PubMed]
113. Acer, E.; Erdogan, H.K.; Igrek, A.; Parlak, H.; Saraçoglu, Z.N.; Bilgin, M. Relationship between diet, atopy, family history, and premature hair graying. *J. Cosmet. Dermatol.* **2019**, *18*, 665–670. [CrossRef]
114. Aggarwal, A.; Srivastava, S.; Agarwal, M.P.; Dwivedi, S. Premature Graying of Hair: An Independent Risk Marker for Coronary Artery Disease in Smokers—A Retrospective Case Control Study. *Ethiop. J. Health Sci.* **2015**, *25*, 123–128. [CrossRef] [PubMed]
115. Mahendiratta, S.; Sarma, P.; Kaur, H.; Kaur, S.; Kaur, H.; Bansal, S.; Prasad, D.; Prajapat, M.; Upadhay, S.; Kumar, S.; et al. Premature graying of hair: Risk factors, co-morbid conditions, pharmacotherapy and reversal—A systematic review and meta-analysis. *Dermatol. Ther.* **2020**, *33*, e13990. [CrossRef]
116. El-Sheikh, A.M.; Elfar, N.N.; Mourad, H.A.; Hewedy, E.S. Relationship between Trace Elements and Premature Hair Graying. *Int. J. Trichology* **2018**, *10*, 278–283. [CrossRef]
117. Choi, J.W.; Lew, B.L.; Sim, W.Y. A Case of Premature Hair Graying Treated with Ferrous Sulfate. *Ann. Dermatol.* **2016**, *28*, 775–776. [CrossRef]
118. Bhat, R.M.; Sharma, R.; Pinto, A.C.; Dandekeri, S.; Martis, J. Epidemiological and investigative study of premature graying of hair in higher secondary and pre-university school children. *Int. J. Trichology* **2013**, *5*, 17–21. [CrossRef]
119. Chakrabarty, S.; Krishnappa, P.G.; Gowda, D.G.; Hiremath, J. Factors Associated with Premature Hair Graying in a Young Indian Population. *Int. J. Trichology* **2016**, *8*, 11–14. [CrossRef]
120. El-Husseiny, R.; Alrgig, N.T.; Abdel Fattah, N.S.A. Epidemiological and biochemical factors (serum ferritin and vitamin D) associated with premature hair graying in Egyptian population. *J. Cosmet. Dermatol.* **2021**, *20*, 1860–1866. [CrossRef]
121. Kiafar, B.; Taheri, A.; Isaac Hashemy, S.; Saki, A.; Mahdeianfar, B.; Taghavi, F.; Vahabi-Amlashi, S. Serum levels of lead and selenium in patients with premature graying of the hair. *J. Cosmet. Dermatol.* **2020**, *19*, 1513–1516. [CrossRef] [PubMed]
122. Skolnik, P.; Eaglstein, W.H.; Ziboh, V.A. Human essential fatty acid deficiency: Treatment by topical application of linoleic acid. *Arch. Dermatol.* **1977**, *113*, 939–941. [CrossRef] [PubMed]
123. Sharma, N.; Dogra, D. Association of Epidemiological and Biochemical Factors with Premature Graying of Hair: A Case-Control Study. *Int. J. Trichology* **2018**, *10*, 211–217. [CrossRef]
124. Sonthalia, S.; Priya, A.; Tobin, D.J. Demographic Characteristics and Association of Serum Vitamin B12, Ferritin and Thyroid Function with Premature Canities in Indian Patients from an Urban Skin Clinic of North India: A Retrospective Analysis of 71 Cases. *Indian J. Dermatol.* **2017**, *62*, 304–308. [CrossRef] [PubMed]
125. Zayed, A.A.; Shahait, A.D.; Ayoub, M.N.; Yousef, A.M. Smokers' hair: Does smoking cause premature hair graying? *Indian Dermatol. Online J.* **2013**, *4*, 90–92. [CrossRef] [PubMed]
126. Sabharwal, R.; Gupta, A.; Moon, N.; Mahendra, A.; Sargaiyan, V.; Gupta, A.; Subudhi, S.K.; Gupta, S. Association between use of tobacco and age on graying of hair. *Niger. J. Surg.* **2014**, *20*, 83–86. [CrossRef]
127. Akin Belli, A.; Etgu, F.; Ozbas Gok, S.; Kara, B.; Dogan, G. Risk Factors for Premature Hair Graying in Young Turkish Adults. *Pediatr. Dermatol.* **2016**, *33*, 438–442. [CrossRef]
128. Osman, M.; McCauley, M.D. Psychiatric shades of grey: Mirtazapine-induced hair discoloration and hair loss. *Ir. J. Psychol. Med.* **2016**, *33*, 175–178. [CrossRef]
129. Richena, M.; Silveira, M.; Rezende, C.A.; Joekes, I. Yellowing and bleaching of grey hair caused by photo and thermal degradation. *J. Photochem. Photobiol. B-Biol.* **2014**, *138*, 172–181. [CrossRef]
130. Zhai, X.; Gong, M.H.; Peng, Y.X.; Yang, D.P. Effects of UV Induced-Photoaging on the Hair Follicle Cycle of C57BL6/J Mice. *Clin. Cosmet. Investig. Dermatol.* **2021**, *14*, 527–539. [CrossRef]
131. Naudin, G.; Bastien, P.; Mezzache, S.; Trehu, E.; Bourokba, N.; Appenzeller, B.M.R.; Soeur, J.; Bornschlögl, T. Human pollution exposure correlates with accelerated ultrastructural degradation of hair fibers. *Proc. Natl. Acad. Sci. USA* **2019**, *116*, 18410–18415. [CrossRef]
132. Bounda, G.A.; Feng, Y.U. Review of clinical studies of *Polygonum multiflorum* Thunb. and its isolated bioactive compounds. *Pharmacogn. Res.* **2015**, *7*, 225–236. [CrossRef]
133. Li, Y.; Han, M.; Lin, P.; He, Y.; Yu, J.; Zhao, R. Hair Growth Promotion Activity and Its Mechanism of *Polygonum multiflorum*. *Evid.-Based Complement. Altern. Med.* **2015**, *2015*, 517901. [CrossRef] [PubMed]
134. Han, M.N.; Lu, J.M.; Zhang, G.Y.; Yu, J.; Zhao, R.H. Mechanistic Studies on the Use of *Polygonum multiflorum* for the Treatment of Hair Graying. *Biomed. Res. Int.* **2015**, *2015*, 651048. [CrossRef]
135. Lin, L.; Ni, B.; Lin, H.; Zhang, M.; Li, X.; Yin, X.; Qu, C.; Ni, J. Traditional usages, botany, phytochemistry, pharmacology and toxicology of *Polygonum multiflorum* Thunb.: A review. *J. Ethnopharmacol.* **2015**, *159*, 158–183. [CrossRef] [PubMed]
136. Ho, T.T.; Murthy, H.N.; Dalawai, D.; Bhat, M.A.; Paek, K.Y.; Park, S.Y. Attributes of *Polygonum multiflorum* to transfigure red biotechnology. *Appl. Microbiol. Biotechnol.* **2019**, *103*, 3317–3326. [CrossRef]
137. Tao, Y.; Zhou, X.; Li, W.; Cai, B. Simultaneous Quantitation of Five Bioactive Ingredients in Raw and Processed *Fallopia multiflora* by Employing UHPLC-Q-TOF-MS. *J. Chromatogr. Sci.* **2019**, *57*, 618–624. [CrossRef]
138. Xia, X.H.; Yuan, Y.Y.; Liu, M. The assessment of the chronic hepatotoxicity induced by Polygoni Multiflori Radix in rats: A pilot study by using untargeted metabolomics method. *J. Ethnopharmacol.* **2017**, *203*, 182–190. [CrossRef]

139. Ling, S.; Xu, J.W. Biological Activities of 2,3,5,4′-Tetrahydroxystilbene-2-O-beta-D-Glucoside in Antiaging and Antiaging-Related Disease Treatments. *Oxid. Med. Cell Longev.* **2016**, *2016*, 4973239. [CrossRef]
140. Lee, S.Y.; Ahn, S.M.; Wang, Z.; Choi, Y.W.; Shin, H.K.; Choi, B.T. Neuroprotective effects of 2,3,5,4′-tetrahydoxystilbene-2-O-beta-D-glucoside from *Polygonum multiflorum* against glutamate-induced oxidative toxicity in HT22 cells. *J. Ethnopharmacol.* **2017**, *195*, 64–70. [CrossRef]
141. Sextius, P.; Betts, R.; Benkhalifa, I.; Commo, S.; Eilstein, J.; Massironi, M.; Wang, P.; Michelet, J.F.; Qiu, J.; Tan, X.; et al. *Polygonum multiflorum* Radix extract protects human foreskin melanocytes from oxidative stress in vitro and potentiates hair follicle pigmentation ex vivo. *Int. J. Cosmet. Sci.* **2017**, *39*, 419–425. [CrossRef] [PubMed]
142. Thang, N.D.; Diep, P.N.; Lien, P.T.H.; Lien, L.T. *Polygonum multiflorum* root extract as a potential candidate for treatment of early graying hair. *J. Adv. Pharm. Technol. Res.* **2017**, *8*, 8–13. [CrossRef] [PubMed]
143. Kim, D.; Kim, H.J.; Jun, H.S. *Polygonum multiflorum* Thunb. Extract Stimulates Melanogenesis by Induction of COX2 Expression through the Activation of p38 MAPK in B16F10 Mouse Melanoma Cells. *Evid.-Based Complement. Altern. Med.* **2020**, *2020*, 7642019. [CrossRef]
144. Jiang, Z.; Xu, J.; Long, M.; Tu, Z.; Yang, G.; He, G. 2,3,5,4′-tetrahydroxystilbene-2-O-beta-D-glucoside (THSG) induces melanogenesis in B16 cells by MAP kinase activation and tyrosinase upregulation. *Life Sci.* **2009**, *85*, 345–350. [CrossRef]
145. Kim, J.; Kim, M.M. The effect of emodin on melanogenesis through the modulation of ERK and MITF signaling pathway. *Nat. Product. Res.* **2022**, *36*, 1084–1088. [CrossRef]
146. Jo, S.J.; Shin, H.; Paik, S.H.; Na, S.J.; Jin, Y.; Park, W.S.; Kim, S.N.; Kwon, O.S. Efficacy and Safety of *Pueraria lobata* Extract in Gray Hair Prevention: A Randomized, Double-Blind, Placebo-Controlled Study. *Ann. Dermatol.* **2013**, *25*, 218–222. [CrossRef]
147. Reichelt, K.V.; Hartmann, B.; Weber, B.; Ley, J.P.; Krammer, G.E.; Engel, K.H. Identification of bisprenylated benzoic acid derivatives from yerba santa (*Eriodictyon* ssp.) using sensory-guided fractionation. *J. Agric. Food Chem.* **2010**, *58*, 1850–1859. [CrossRef]
148. Yang, J.; Liang, Q.; Wang, M.; Jeffries, C.; Smithson, D.; Tu, Y.; Boulos, N.; Jacob, M.R.; Shelat, A.A.; Wu, Y.S.; et al. UPLC-MS-ELSD-PDA as a Powerful Dereplication Tool to Facilitate Compound Identification from Small-Molecule Natural Product Libraries. *J. Nat. Prod.* **2014**, *77*, 902–909. [CrossRef]
149. Walker, J.; Reichelt, K.V.; Obst, K.; Widder, S.; Hans, J.; Krammer, G.E.; Ley, J.P.; Somoza, V. Identification of an anti-inflammatory potential of *Eriodictyon angustifolium* compounds in human gingival fibroblasts. *Food Funct.* **2016**, *7*, 3046–3055. [CrossRef]
150. Taguchi, N.; Homma, T.; Aoki, H.; Kunisada, T. Dietary *Eriodictyon angustifolium* Tea Supports Prevention of Hair Graying by Reducing DNA Damage in CD34+ Hair Follicular Keratinocyte Stem Cells. *Biol. Pharm. Bull.* **2020**, *43*, 1451–1454. [CrossRef] [PubMed]
151. Taguchi, N.; Hata, T.; Kamiya, E.; Kobayashi, A.; Aoki, H.; Kunisada, T. Reduction in human hair graying by sterubin, an active flavonoid of *Eriodictyon angustifolium*. *J. Dermatol. Sci.* **2018**, *92*, 286–289. [CrossRef] [PubMed]
152. Taguchi, N.; Hata, T.; Kamiya, E.; Homma, T.; Kobayashi, A.; Aoki, H.; Kunisada, T. *Eriodictyon angustifolium* extract, but not Eriodictyon californicum extract, reduces human hair greying. *Int. J. Cosmet. Sci.* **2020**, *42*, 336–345. [CrossRef] [PubMed]
153. Taguchi, N.; Kitai, R.; Ando, T.; Nishimura, T.; Aoki, H.; Kunisada, T. Protective Effect of Hydroxygenkwanin against Hair Graying Induced by X-Ray Irradiation and Repetitive Plucking. *JID Innov.* **2022**, *2*, 100121. [CrossRef]
154. Aoki, H.; Yamada, Y.; Hara, A.; Kunisada, T. Two distinct types of mouse melanocyte: Differential signaling requirement for the maintenance of non-cutaneous and dermal versus epidermal melanocytes. *Development* **2009**, *136*, 2511–2521. [CrossRef] [PubMed]
155. Itoh, T.; Furuichi, Y. Hot-water extracts from adzuki beans (*Vigna angularis*) stimulate not only melanogenesis in cultured mouse B16 melanoma cells but also pigmentation of hair color in C3H mice. *Biosci. Biotechnol. Biochem.* **2005**, *69*, 873–882. [CrossRef]
156. Zhao, P.J.; Park, N.H.; Alam, M.B.; Lee, S.H. Fuzhuan Brick Tea Boosts Melanogenesis and Prevents Hair Graying through Reduction of Oxidative Stress via NRF2-HO-1 Signaling. *Antioxidants* **2022**, *11*, 599. [CrossRef]
157. Liu, X.; Lv, X.; Ji, T.; Hu, H.; Chang, L. *Gynostemma pentaphyllum* Makino extract induces hair growth and exhibits an anti-graying effect via multiple mechanisms. *J. Cosmet. Dermatol.* **2024**, *23*, 648–657. [CrossRef] [PubMed]
158. de la Vega, M.R.; Zhang, D.D.; Wondrak, G.T. Topical Bixin Confers NRF2-Dependent Protection Against Photodamage and Hair Graying in Mouse Skin. *Front. Pharmacol.* **2018**, *9*, 287. [CrossRef] [PubMed]
159. Li, X.X.; Shi, R.L.; Yan, L.C.; Chu, W.W.; Sun, R.S.; Zheng, B.K.; Wang, S.; Tan, H.; Wang, X.S.; Gao, Y. Natural product rhynchophylline prevents stress-induced hair graying by preserving melanocyte stem cells via the β2 adrenergic pathway suppression. *Nat. Prod. Bioprospecting* **2023**, *13*, 54. [CrossRef]
160. Wang, Y.; Yao, X.; Shen, H.; Zhao, R.; Li, Z.; Shen, X.; Wang, F.; Chen, K.; Zhou, Y.; Li, B.; et al. Nutritional Composition, Efficacy, and Processing of *Vigna angularis* (Adzuki Bean) for the Human Diet: An Overview. *Molecules* **2022**, *27*, 6079. [CrossRef]
161. Chen, G.; Peng, Y.; Xie, M.; Xu, W.; Chen, C.; Zeng, X.; Liu, Z. A critical review of Fuzhuan brick tea: Processing, chemical constituents, health benefits and potential risk. *Crit. Rev. Food Sci. Nutr.* **2023**, *63*, 5447–5464. [CrossRef] [PubMed]
162. Tao, S.; Park, S.L.; Rojo de la Vega, M.; Zhang, D.D.; Wondrak, G.T. Systemic administration of the apocarotenoid bixin protects skin against solar UV-induced damage through activation of NRF2. *Free Radic. Biol. Med.* **2015**, *89*, 690–700. [CrossRef] [PubMed]
163. Qin, N.; Lu, X.; Liu, Y.; Qiao, Y.; Qu, W.; Feng, F.; Sun, H. Recent research progress of *Uncaria* spp. based on alkaloids: Phytochemistry, pharmacology and structural chemistry. *Eur. J. Med. Chem.* **2021**, *210*, 112960. [CrossRef] [PubMed]
164. Tobin, D.J. Aging of the hair follicle pigmentation system. *Int. J. Trichology* **2009**, *1*, 83–93. [CrossRef] [PubMed]

165. Gelmi, M.C.; Houtzagers, L.E.; Strub, T.; Krossa, I.; Jager, M.J. MITF in Normal Melanocytes, Cutaneous and Uveal Melanoma: A Delicate Balance. *Int. J. Mol. Sci.* **2022**, *23*, 6001. [CrossRef]
166. Heilmann-Heimbach, S.; Hochfeld, L.M.; Henne, S.K.; Nothen, M.M. Hormonal regulation in male androgenetic alopecia-Sex hormones and beyond: Evidence from recent genetic studies. *Exp. Dermatol.* **2020**, *29*, 814–827. [CrossRef]
167. Wellbrock, C.; Arozarena, I. Microphthalmia-associated transcription factor in melanoma development and MAP-kinase pathway targeted therapy. *Pigment Cell Melanoma Res.* **2015**, *28*, 390–406. [CrossRef]
168. Lee, A.R.; Lim, J.; Lim, J.S. Emerging roles of MITF as a crucial regulator of immunity. *Exp. Mol. Med.* **2024**, *56*, 311–318. [CrossRef]
169. Kim, J.H.; Seok, J.K.; Kim, Y.M.; Boo, Y.C. Identification of small peptides and glycinamide that inhibit melanin synthesis using a positional scanning synthetic peptide combinatorial library. *Br. J. Dermatol.* **2019**, *181*, 128–137. [CrossRef]
170. Mansky, K.C.; Sankar, U.; Han, J.H.; Ostrowski, M.C. Microphthalmia transcription factor is a target of the p38 MAPK pathway in response to receptor activator of NF-κB ligand signaling. *J. Biol. Chem.* **2002**, *277*, 11077–11083. [CrossRef]
171. Zhou, J.; Song, J.; Ping, F.F.; Shang, J. Enhancement of the p38 MAPK and PKA signaling pathways is associated with the pro-melanogenic activity of Interleukin 33 in primary melanocytes. *J. Dermatol. Sci.* **2014**, *73*, 110–116. [CrossRef]
172. Kim, J.H.; Hong, A.R.; Kim, Y.H.; Yoo, H.; Kang, S.W.; Chang, S.E.; Song, Y. JNK suppresses melanogenesis by interfering with CREB-regulated transcription coactivator 3-dependent MITF expression. *Theranostics* **2020**, *10*, 4017–4029. [CrossRef] [PubMed]
173. Tonelli, C.; Chio, I.I.C.; Tuveson, D.A. Transcriptional Regulation by Nrf2. *Antioxid. Redox Signal* **2018**, *29*, 1727–1745. [CrossRef] [PubMed]
174. Boo, Y.C. Natural Nrf2 Modulators for Skin Protection. *Antioxidants* **2020**, *9*, 812. [CrossRef]
175. Kwack, M.H.; Sung, Y.K.; Chung, E.J.; Im, S.U.; Ahn, J.S.; Kim, M.K.; Kim, J.C. Dihydrotestosterone-inducible dickkopf 1 from balding dermal papilla cells causes apoptosis in follicular keratinocytes. *J. Investig. Dermatol.* **2008**, *128*, 262–269. [CrossRef]
176. Papukashvili, D.; Rcheulishvili, N.; Liu, C.; Xie, F.F.; Tyagi, D.; He, Y.J.; Wang, P.G. Perspectives on miRNAs Targeting DKK1 for Developing Hair Regeneration Therapy. *Cells* **2021**, *10*, 2957. [CrossRef] [PubMed]
177. Kong, Y.N.; Liu, Y.S.; Pan, L.L.; Cheng, B.A.; Liu, H.W. Norepinephrine Regulates Keratinocyte Proliferation to Promote the Growth of Hair Follicles. *Cells Tissues Organs* **2015**, *201*, 423–435. [CrossRef]
178. De Tollenaere, M.; Chapuis, E.; Auriol, P.; Auriol, D.; Scandolera, A.; Reynaud, R. Global Repigmentation Strategy of Grey Hair Follicles by Targeting Oxidative Stress and Stem Cells Protection. *Appl. Sci.* **2021**, *11*, 1533. [CrossRef]
179. Park, H.J.; Zhang, N.; Park, D.K. Topical application of Polygonum multiflorum extract induces hair growth of resting hair follicles through upregulating Shh and β-catenin expression in C57BL/6 mice. *J. Ethnopharmacol.* **2011**, *135*, 369–375. [CrossRef]
180. Yu, H.S.; Wang, L.L.; He, Y.; Han, L.F.; Ding, H.; Song, X.B.; Gao, X.M.; Yun, N.R.; Li, Z. Advances in the Study of the Potential Hepatotoxic Components and Mechanism of *Polygonum multiflorum*. *Evid.-Based Complement. Altern. Med.* **2020**, *2020*, 6489648. [CrossRef]
181. Teka, T.; Wang, L.M.; Gao, J.; Mou, J.J.; Pan, G.X.; Yu, H.Y.; Gao, X.M.; Han, L.F. *Polygonum multiflorum*: Recent updates on newly isolated compounds, potential hepatotoxic compounds and their mechanisms. *J. Ethnopharmacol.* **2021**, *271*, 113864. [CrossRef] [PubMed]

applied sciences

MDPI

Review

Recent Studies on the Healing Properties of Eicosapentaenoic Acid

Maria Dospra [1], Panagoula Pavlou [1,2], Spyridon Papageorgiou [1,2] and Athanasia Varvaresou [1,2,*]

1 Sector of Aesthetics and Cosmetic Science, Department of Biomedical Sciences, School of Health and Care Sciences, University of West Attica, 28 Ag. Spyridonos Str., Panepistimioupolis Egaleo Park, 12243 Athens, Greece; bisc18678224@uniwa.gr (M.D.); ppavlou@uniwa.gr (P.P.); spapage@uniwa.gr (S.P.)

2 Laboratory of Chemistry-Biochemistry-Cosmetic Science, Department of Biomedical Sciences, University of West Attica, 28 Ag. Spyridonos Str., Panepistimioupolis Egaleo Park, 12243 Athens, Greece

* Correspondence: avarvares@uniwa.gr

Abstract: Patients with diabetes, the elderly, and those who have certain inherited conditions are particularly vulnerable to atypical wound healing with lingering repercussions. Remarkably, the current remedies are insufficient since, despite the plethora of wound healing options, only limited efficacy is observed. This review is a bibliographic survey on eicosapentaenoic acid and its healing effects. It has been investigated in terms of its source of origin, structure, physico-chemical properties, and studies where healing action is demonstrated. Fatty acids, found in all layers of the skin, modify cell function and the synthesis of eicosanoids, reactive oxygen species, and cytokines, which affects skin structure and immunological condition. As a result, fatty acids influence both the inflammatory response and the wound-healing process. EPA is one of the dietary lipids that has a variety of health advantages. It functions in anti-inflammatory processes and the firmness of cell membranes and is integrated into numerous bodily parts. EPA has a crucial role in healthy fetal development and aging. It is a precursor to numerous metabolites that are powerful lipid mediators and are regarded by many researchers as being helpful in the treatment or prevention of several disorders. EPA supplement is used after surgery to lessen infections, accelerate wound healing, and speed up recovery, although, according to other researchers, the oral administration of omega-3 fatty acids, particularly the DHA and EPA combination, significantly slows down the healing of wounds and disrupts the structure of collagen through several mechanisms. The controversy in the reported literature is discussed and new technologies useful for the improvement of the wound healing process are also reported.

Keywords: eicosapentaenoic acid; wound healing; diabetes; skin; oral administration

1. Introduction

The management of acute and chronic wounds has a significant financial, clinical, and social impact on healthcare and affects the living conditions of over 2.5% of the population. Understanding the wound's etiology, chronicity, healing mechanism, and biology is necessary for providing ideal wound care [1]. Despite the numerous treatments already established for both chronic and acute wounds, their management continues to constitute an urgent and burdensome issue in modern times, necessitating the development of novel therapies.

Injuries have always been difficult to treat and continue to put a tremendous strain on the medical system. Over 8 million Americans suffered from wounds in 2014, with an estimated USD 30 billion in medical expenses. High-risk comorbidities rise in tandem with an aging and obese population, with the wound closure product market expected to reach USD 21.4 billion in 2022 with a compound annual growth rate of 4.15% from 2023 to 2030 [2].

The goal of commercially accessible biomaterials for wound healing is usually symptom treatment (infection, pressure reduction, scarring, moisture balance, fluid exudation, etc.). On the other hand, sophisticated biomaterials for wound healing are being created to offer biophysical cues inspired by extracellular matrix (ECM) and immune response modulation for sufficient inflammation resolution. These treatments may combine the use of pharmaceutical and biological product therapies, and they are usually designed as injectable or biomaterial-based delivery systems. Basic research has demonstrated how biophysical signals can be included into biomaterials to regulate the activity of cells. Systems of delivery based on biomaterials, such as hydrogels, can offer stimuli-responsive and sustained-release characteristics. These ideas could further encourage patient adherence to novel treatments and get around the drawbacks and hazards of systemic delivery [2,3].

Numerous studies have suggested that fish oil supplements may lessen the severity of a number of skin conditions, including melanogenesis, dermatitis, allergies, photoaging, and skin cancer. The correlation between fish oil and skin protection and homeostasis has garnered significant attention, particularly with the omega-3 polyunsaturated fatty acids (PUFAs), namely docosahexaenoic acid (DHA) and eicosapentaenoic acid (EPA). The two main ways that polyunsaturated fatty acids (PUFAs) reduce cutaneous inflammation are by competing with inflammatory arachidonic acid and by preventing the synthesis of proinflammatory eicosanoid molecules [4].

PUFAs in fish oil, on the other hand, may function as regulators to change the synthesis and activity of cytokines for promoting wound healing. Proinflammatory cytokines are mostly responsible for the initiation and amplification of the cellular and molecular processes that occur during the inflammatory phase of wound healing. PUFAs have the ability to control the production and function of cytokines. It has been demonstrated that these fatty acids are essential for the formation of cell membranes and anabolic processes that occur during skin tissue regeneration. The local inflammatory response at wound sites may be modulated or enhanced by omega-3 and omega-6 polyunsaturated fatty acids, which could hasten the healing process [5]. Using the most recent research, this study attempts to provide a bibliographic survey on eicosapentaenoic acid and its therapeutic properties.

Methodology: The academic literature related to this specific bioactive compound and its potential wound healing effect were studied. The literature search was based on scientific databases, including Scopus, PubMed, Science Direct, and Google Scholar, mainly from 2017 to 2024. A total of 66 articles, including research papers and book chapters, were downloaded, of which 32 more recent articles on closer relevance to the topic were selected for writing the review. The keywords used were eicosapentaenoic acid, wound healing, skin, diabetes, and in vivo.

We included pre-clinical and clinical studies and we avoided including articles that were more focused on the biochemical mechanism of action than the results of the administration of EPA.

We think it is worth searching the literature during the last seven years (2017–2024), since we found enough articles relevant to the clinical application of ERA, since the support of wound healing and many dermatological problems by supplements is a trend of the recent years.

2. Skin and Its Healing Mechanisms

The skin, which covers 1.85 m^2 of surface area on average, is the biggest organ in the human body. It is the natural barrier against pathogenic microbes, extreme temperature, mechanical damage, toxic agents, and UV radiation [6]. According to their impact and underlying causes, skin wounds are divided into acute and chronic categories [7]. In contrast to chronic wounds, which are challenging to heal, acute wounds typically heal in 4–12 weeks with the recovery of the functional and anatomical integrity of the skin. Chronic wounds take longer to heal, a fact that increases the risk of infection and hinders the full and effective restoration of skin integrity. Adults usually exhibit scarring rather than total tissue regeneration as a result of persistent or deep wounds.

The epidermis, dermis, and subcutaneous adipose tissue, which run from the outer to the inner skin layer, must cooperate to perform particular sequences to foster healing when the skin is wounded. There are four stages of healing, which include hemostasis (I);

inflammation (II); proliferation (III); and granulation tissue creation, re-epithelialization, and (IV) remodeling, which take place in a certain order or overlap.

Trauma causes the wounded blood vessels to constrict first, followed by the activation of platelets to form a clot. The clot seals the wound temporarily, preventing the spread of germs, and serves as a scaffold for arriving inflammatory cells [8]. Damaged epithelial and endothelial cells, along with the clot, generate several chemotactic substances that draw in inflammatory cells to commence the following phase of healing. After hemostasis, the inflammatory phase follows, where immune cells travel to the site of the wound as a reaction to damage signaling factors to ward off microbial infection [7].

At the location of the wound, there is an inflow of various inflammatory cells, that consequently produce a range of mediators and cytokines that facilitate angiogenesis and re-epithelialization. Hemostasis, chemotaxis, and enhanced vascular permeability are the characteristics of the inflammatory phase, which limit additional damage, seal the wound, cleanse cellular debris and microorganisms, and accelerate cell migration [9]. The clot thrombus is instantly inhabited by neutrophils as the first line of resistance against pathogens. Within 48–96 h of injury, monocytes are drawn in to change into tissue-activated macrophages in the damaged area. To fight self- and foreign antigens, the adaptive immune system, which consists of Langerhans cells, dermal dendritic cells, and T cells, is also activated [8].

Acute phase signals are first presented to skin cells, which are then recognized by toll-like receptors that start and maintain inflammation. To make it to the injury site, leukocytes, most notably neutrophils, migrate along a gradient of chemokines that keeps growing. Additionally, neutrophils release a variety of pro-inflammatory cytokines which boost the inflammatory response [6]. Apoptosis, or necrosis, is followed by macrophage engulfment, or efferocytosis, to start the process of clearing out neutrophils. Neutrophils must be eliminated quickly since their removal initiates the process of inflammation resolution [8]. The adaptive immune system includes activated regulatory T cells. Regulatory T cells, in addition to leukocytes, have the ability to reduce tissue inflammation by lowering the production of interferon-γ as well as the accumulation of proinflammatory macrophages. Macrophages are one of the significant factors in the progression from inflammation phase to proliferation. Pathogen-specific and damage-related molecular patterns activate macrophages. This causes their differentiation into the M1 subset, which is linked to phagocytic activity, scavenging, and also the synthesis of pro-inflammatory mediators, in the early stages of wound repair. Later, M1 changes into the M2 subset, displaying reparative characteristics. M2 macrophages play a role in fibroblast proliferation, the creation of the extracellular matrix (ECM), the synthesis of anti-inflammatory mediators, and angiogenic procedures. M2 macrophages operate as a sort of cleanup crew by phagocytosing neutrophils, bacteria, and cellular debris, eliminating additional damage to the wound site late in the healing process [6]. The inflammatory phase, which typically lasts from one to four days following injury, clears the wound of debris, and performs significant phagocytosis to prepare it for regeneration. Various cells, including fibroblasts, keratinocytes, and endothelial cells, are recruited by growth factors and cytokines released by immune system cells throughout the inflammatory phase to kick off the proliferation phase [7]. This stage's duration might be several weeks [9]. Fibroblasts, the ECM creation for granulation tissue formation, the proliferation and migration of endothelial cells (angiogenesis), and the repair of the epidermal barrier by means of wound re-epithelialization all participate in the process of skin tissue remodeling.

The ECM components created by fibroblasts' proliferation and formation of granulation tissue serve as a substrate for the development of new blood vessels [7]. Macrophages, fibroblasts, blood vessels, and a loose extracellular matrix consisting of type I collagen, glycoprotein, fibronectin, and hyaluronic acid compose the granulation tissue. Re-epithelialization caused by keratinocyte differentiation and proliferation covers the wound with overlying epidermal tissue. Following this stage is the remodeling or maturation phase, during which the granulation tissue is replaced by type I collagen and various skin components, i.e., sweat glands, sebaceous glands, and hair follicles [7]. To provide the skin with better tensile strength, during this phase, which starts around the third week and can last up to 12 months [9], fibroblasts drop into myofibroblasts and type III collagen transforms into type I collagen [7]. The tensile strength reaches its peak after 11 to 14 weeks. Only around 80% of the original tensile strength remains in the resulting scar, which means it will never regain its full original tensile strength [9]. In the case of acute wounds, wound healing occurs when all of these stages have been completed. Chronic wounds, however, do not go through all of these stages, and the healing process is impeded [10].

2.1. Sources of Eicosapentaenoic Acid, Structure, and Physico-Chemical Properties

The polyunsaturated fatty acid (PUFA) class of omega-3 fatty acids includes eicosapentaenoic acid (EPA) (Figure 1). Humans can consume EPA through their diets by eating fatty fish, different kinds of edible algae, or by supplementation with fish oil or algal oil. Cod and plaice oil are the fish oils with the greatest concentrations of EPA (Table 1). Alpha linolenic acid (ALA), which is absorbed by the body, is partially converted into EPA. Humans require a sufficient amount of ALA because it is a necessary fatty acid. However, the efficiency of converting ALA to EPA is significantly less than the absorption of EPA from food intake [11].

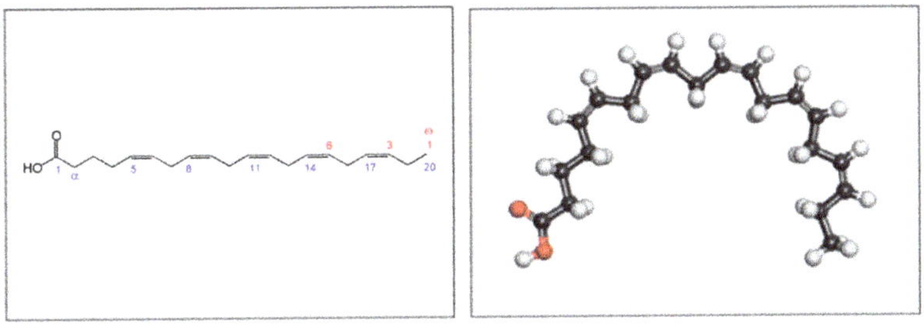

Figure 1. Two-dimensional and three-dimensional molecular structure of EPA [12].

Table 1. EPA SOURCES. Foods and the grams of EPA they contain per 100 g Saini and Keum 2018 [11].

Foods	EPA [g/100 g]
Salmon	13.20
Sardine	10.14
Cod liver	9.90
Herring	6.27
Cooked beef	0.15

The membranes of cells contain EPA. EPA forms eicosanoids during an inflammatory reaction by being metabolized by the cyclooxygenase and lipoxygenase enzymes. Because EPA is used to metabolize docosahexaenoic acid (DHA) as its precursor, which requires additional metabolic energy to produce, it is important to ensure that there is an adequate amount of EPA in the diet. Eicosanoids produced from EPA are generally strong inducing agents of vasoconstriction, inflammation, and blood coagulation [12].

Thromboxane (TX)A_3 derived from EPA is weakly proaggregatory when compared with the arachidonic acid (AA)-derived TXA_2. Also, the influence of pro-inflammatory PGE_2 and the leukotriene (LT)B_4, which is a chemoattractant and activator of neutrophils, is affected by the presence of EPA. Although PGE_3 has a similar action as PGE_2, it is produced with low efficiency. LTB_5 has little inflammatory activity when compared with LTB_4.

Of particular interest are findings on the highly inducible COX-2 which is involved in the overproduction of prostaglandins in inflammatory sites. EPA greatly decreased the interleukin-(IL-)1-induced COX-2 expression in endothelial cells. Although it is well documented that ω-3 fatty acids suppress cytokine production such as tumor necrosis factor-(TNF-)α and IL-1, the underlying mechanisms still remain unclear. It appears that interfering with the synthesis of TXA_2 and PGE_2 also affects cytokine production (Figure 2) [13].

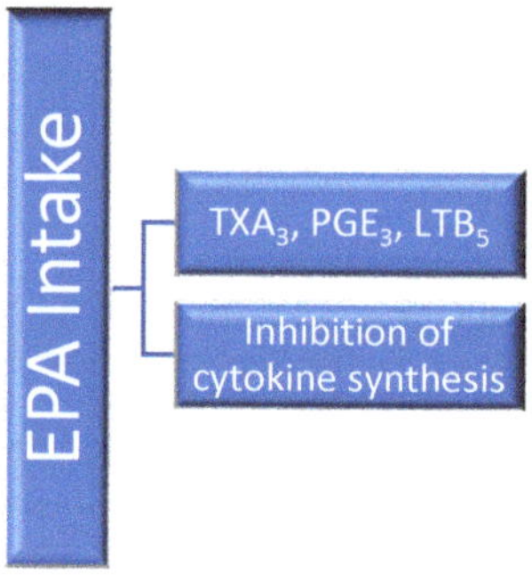

Figure 2. Schematic representation of the effects of eicosapentaenoic acid (EPA). EPA attenuates some of the actions of arachidonic acid (AA) particularly via the synthesis of TXA_3, PGE_3, and LTB_5 [13].

2.2. Recent Studies on Skin

EPA has a healing effect, as proven by Eugenia Sofrona and her team. In that study, lipophilic and hydrophilic extracts of the marine isopod *Ceratothoa oestroides* were produced, and their ability to promote wound healing after being applied topically to wounds induced on SKH-hr1 hairless mice was assessed. Measurements of skin biophysical parameters such as TEWL, hydration, elasticity, and skin thickness and histological examination were primarily used to evaluate wound healing daily. Using an EPA-rich fraction to treat wounds resulted in the most activity, according to the findings of the preliminary evaluation of the various extracts. Mice's wounds entirely healed after being treated locally with the bioactive fraction, and the skin completely returned to normal without any signs of inflammation [14].

Recently, Vitsos A. et al., studied the healing efficacy of composite dressings containing Pycnogenol™ and/or *Ceratothoa oestroides* extracts, with the goal of improving their efficiency and stability while applying a low-temperature vacuum technique. Surgical wounds were inflicted on SKH-hr2 hairless mice. The assessment of wound healing was mainly conducted by clinical and histological examination, with an Antera 3D camera and biophysical measurements. Dressings remained constant and did not affect the *C. oestroides* extract's medicinal qualities. As a standard product, all interventions were contrasted with the *C. oestroides* ointment. By the fifteenth day, the majority of the wounds treated with the reference formulation and the *C. oestrodes* dressing had closed, with histological scores of 6.5 and 7, respectively. Considering the formulations that contained *C. oestroides* olive oil extract, the results showed that, in comparison to the control, the extract improved the healing process [15].

Karoud et al., reported that the usually wasted hake heads from *M. merluccius* might be used as a unique raw material for the production of oil. This fish oil is a good source of n-6 PUFAs, including arachidonic acid, and n-3 PUFAs, especially EPA and DHA. This study presented data on the anti-inflammatory and antioxidant activities of hake head oil (HHO). Using a round seal, a uniform wound area measuring 1 cm in diameter was created on the mouse's previously cleaned and shaved back. Three mouse groups, each with six mice, were given the appropriate treatment: Group I received sterile saline treatment and was assigned as the negative control. Group III received a topical application of 300 mg/kg of hake head oil (HHO), and Group II received a topical application of 0.13 mg/mm^2 of CICAFLORA® (comparison medication), designated as the reference group. Every two days, HHO and the reference medication, CICAFLORA®, were administered until the wound had healed completely. When compared to the negative control, the HHO- and CICAFLORA®-treated groups showed appreciable evolutions in wound contraction rates

($p < 0.05$). The HHO-treated group showed the strongest healing contraction rate during the cicatrization process as compared to the CICAFLORA®-treated group. In fact, on the sixth day, they observed a reduction in the wound area for HHO and CICAFLORA® of 55.66% and 67.89%, respectively. In actuality, the HHO-treated group showed 100% wound healing at the conclusion of the experiment, compared to only 80% for the CICAFLORA®-treated group. The abundance of n-3 and n-6 polyunsaturated fats in this oil, along with its antioxidant and anti-inflammatory qualities, may be contributing factors to HHO's potential to heal wounds [16].

Ontoria-Oviedo et al., examined the impact on wound healing of a marine lipid concentrate developed from anchovy and sardine oil that is high in SPMs (specialized lipid mediators), EPA, and precursor substance DHA (docosahexaenoic acid), demonstrating the healing power of EPA. The marine lipid concentrates enhanced wound closure in vitro, decreased the number of proinflammatory macrophages (M1), and demonstrated biocompatibility with keratinocytes and fibroblasts. In this study, daily administration of EPA-rich marine oil to open wounds in db/db mice accelerated the healing process by reducing inflammation, promoting neoangiogenesis, and polarizing M1/M2 macrophages. This process improved wound closure [17].

In a recent study, the effect of lipid derivatives from EPA and DHA on burn wound healing was investigated. In Yorkshire pigs, after partial and full thickness burn wounds were created, punch biopsies were collected at various time points during treatment and analyzed for measurements of approximately 45 EPA and DHA derivatives. The results showed nine lipid mediators of EPA and DHA to be elevated on the seventh day of treatment in full-thickness burns, thus suggesting the healing effect of these fatty acids on wounds [18].

2.3. Oral Consumption

As EPA belongs to the omega-3 fatty acids, which partake in many functions of the body, and has recognized beneficial properties, it is reasonable to see why it is a component of many food supplements. EPA is available in fish oil capsules, along with DHA [docosahexaenoic acid]. The majority of commercially available dietary supplements are made from fish oil and are often offered as triglycerides, ethyl ester, or phospholipid EPA.

McDaniel J. C. and his team, taking advantage of the healing effect of EPA, compared the efficacy of EPA + DHA fish oil capsule therapy with a placebo in reducing polymorphonuclear leukocyte activation in chronic venous leg ulcers. In this study, 40 patients with chronic venous leg ulcers were randomly assigned to either placebo [Control Group] or EPA + DHA supplementation. Polymorphonuclear leukocyte (CD15) and activated polymorphonuclear leukocyte (CD66b) markers and protease levels were measured on days 0, 28, and 56 of treatment. The collective findings concluded that oral EPA + DHA treatment can modulate the activity of polymorphonuclear leukocytes and facilitate the healing of chronic venous leg ulcers [19].

The same team of researchers also investigated a novel oral therapy containing eicosapentaenoic acid (EPA) and docosahexaenoic acid (DHA), two bioactive components of fish oil, to target and reduce the elevated amounts of activated polymorphonuclear leukocytes in the site of the wound that keep chronic venous leg ulcers "enslaved" in a chronic inflammatory state. In this clinical research, 248 adults over 55 years of age with chronic venous leg ulcers were divided into two groups and randomly assigned to be given either EPA + DHA (1.87 g/day EPA + 1.0 g/day DHA) or placebo daily for 12 weeks. Then, during the treatment, they examined factors such as reduction in the wound area, active polymorphonuclears in the blood, lipid metabolites, etc., in the patients. The results provided new evidence on the efficacy of EPA + DHA treatment, which may facilitate healing and thus be a novel adjunctive therapy for chronic venous leg ulcers in the aging population [20].

Soleimani et al. examined the benefits of flaxseed oil supplementation with additional omega-3 fatty acids on wound healing and metabolic status in patients with DFU focused on healing such ulcers. Patients for the study were divided at random into two groups and given either 1000 mg of flaxseed supplements containing omega-3 fatty acids [EPA and DHA] or a placebo twice daily for 12 weeks. When compared to the placebo group, the supplement group exhibited a substantial decrease in dimensional extent of the ulcer. Significant decreases in insulin resistance and blood insulin levels were seen. Overall, measurements of diabetic foot ulcer size and indicators of the metabolism of insulin benefited from flaxseed supplementation with omega-3 fatty acids [21].

Previous research has indicated that the decreased production of growth factors such as platelet-derived growth factor (PDGF), vascular endothelial growth factor (VEGF), keratinocyte growth factor (KGF), and excess protease activity may be associated with potential etiologies for impaired wound healing in diabetes. A crucial part of diabetic wound healing is played by mast cells. Mast cell degranulation can be influenced by omega-3 fatty acids in a dose-dependent manner. Babaei et al. investigated the association between the presence and degranulation of mast cells in the diabetic wound region and omega-3 fatty acids. After shaving their backs, the rats had an extensive wound (20 mm × 20 mm) that measured and reached the level of the panniculus carnosus muscle. Omegaven (Fluka/Sigma-Aldrich®, St. Louis, MO, USA) was administered intraperitoneally to the experimental group ($n = 35$) twice a day at a dose of 1 mL per rat (100 mL of Omegaven contains 2.82 g of EPA and 3.09 g of DHA). Before the trial began, administration was initiated intraperitoneally at a dose of 1 mL once daily. This was continued until the conclusion of each analysis (biomechanical test and histologic examination). When comparing the wound area of experimental and control groups on day 7 and the quantity of grade three mast cells on days 3 and 5, it was shown to be significantly reduced by omega-3 fatty acids. It was also found that the wound strength significantly increased in the experimental group on day 15 [22].

Notwithstanding the generally acknowledged health advantages of omega-3 fatty acids, particular research has indicated unfavorable outcomes of these substances in the process of wound healing. Burger et al. (2019) examined how oral eicosapentaenoic acid (EPA)-rich oil delivery affects mice's ability to repair wounds. Following four weeks of supplementation with 2 g/kg of body weight with EPA-rich oil, mice's serum concentrations of docosahexaenoic acid (DHA; 22:6ω-3) (33%) and EPA (20:5ω-3) were higher than those of control mice. The mice who were fed EPA also had skin that contained omega-3 fatty acids. Compared to control mice, mice that received EPA-rich oil showed a delay in the healing process at the 3- and 7-day marks after wounding; however, this delay did not affect the overall time needed for wound closure. Following EPA-rich oil supplementation, collagen reorganization—which affects the integrity of the wound tissue—was compromised. These effects were linked to increased amounts of interleukin-10 (IL-10) in tissue during the early phases of wound healing, as well as twice as many M2 macrophages compared to control rats. Wound closure and collagen organization returned to normal in the absence of IL-10 (IL-$10^{-/-}$ mice), indicating that the harmful effects of EPA-rich oil supplementation were caused by the overproduction of IL-10. To sum up, the quality of wound healing was compromised by oral administration of EPA-rich oil, but the wound closure time remained unaffected. This is probably because the anti-inflammatory cytokine IL-10 was elevated [23].

Candreva et al. (2019) employed two distinct methods. Two types of mice were used: (1) FAT-1 transgenic mice that were able to produce ω-3 FA naturally; and (2) wild-type (WT) mice that were given DHA-enriched fish oil orally. Compared to WT mice, FAT-1 mice exhibited elevated levels of ω-3 FA, primarily docosahexaenoic acid (DHA), in both their systemic (serum) and local (skin tissue) samples. Three days following the formation of the wound, the cutaneous wound tissue of FAT-1 mice exhibited decreased IL-10 levels, increased myeloperoxidase (MPO) activity, and higher levels of CXCL-1 and CXCL-2. The presence of inflammatory cells, edema, and an elevated TNF-α concentration all contributed to the maintenance of inflammation. Isolated from FAT-1 mice, neutrophils and macrophages also produced lower levels of IL-10 and higher levels of TNF-α. On the final day of analysis, which was 14 days after the wound, the mice's wounds were six times larger than those of the WT group due to a delayed wound closure. This was linked to structural features in the restored tissue as well as improper collagen fiber orientation. In a similar vein, the DHA group saw a late inflammatory phase delay. On the third day following skin injury, this group exhibited elevated TNF-α content and CD45+F4/80+ cells along with reduced concentrations of ω-6 FA metabolites and increased concentrations of key metabolites generated from ω-3, such as 18-HEPE. In conclusion, both the FAT-1 and DHA groups' increased DHA content retarded the resolution of inflammation and decreased the quality of the skin tissue that healed [24].

Burger et al. [25] showed that EPA supplementation has a negative effect on diabetic mice's ability to heal wounds due to decreased collagen gene expression, poor collagen organization, and a resulting decrease in the resistance of the repaired tissue, mechanistically, via activating PPAR-γ (peroxisome proliferator-activated receptors) [19]. EPA supplementation increased neutrophils' production of IL-10. Simultaneously, EPA and IL-10 decreased collagen expression and fibroblasts' synthesis of TGF-β. The discovery that PPAR-γ blockage in vivo prevented the EPA-induced decrease in collagen organization could be the most significant. In more detail, EPA-rich oil reduced the ω-6/ω-3 ratio by increasing the incorporation of ω-3 and decreasing the ω-6 fatty acids, according to the gas chromatography study of serum and skin. Ten days after wounding, EPA induced neutrophils in the wound to produce more IL-10, decreased collagen deposition, and eventually postponed wound closure and compromised the quality of the recovered tissue. This effect was dependent on PPAR-γ. In vitro fibroblasts produced less collagen when exposed to EPA and IL-10. In vivo, topical PPAR-γ-blockade in diabetic mice mitigated the detrimental effects of EPA on collagen organization and wound closure. In diabetic animals given the PPAR-γ blocker topically, neutrophils produced less IL-10. These findings demonstrated how oral EPA-rich oil supplementation hinders the healing of skin wounds in diabetics by influencing both inflammatory and non-inflammatory cells (Figure 3) [25].

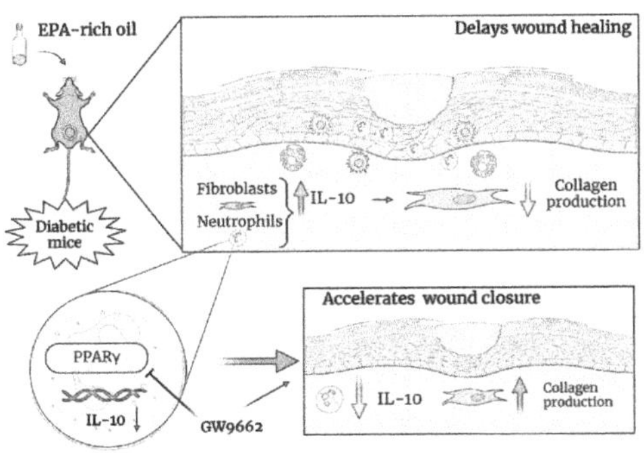

Figure 3. Oral EPA-rich oil supplementation hinders the healing of skin wounds in diabetics [25].

3. Discussion

EPA is widely known for the management or prevention of various heart-related diseases, such as blocked heart arteries [coronary artery disease] and heart attacks. It also reduces triglyceride levels in those with very elevated levels [26]. It finds application along with the appropriate medicines, in the treatment of schizophrenia, personality disorders, Alzheimer's disease, depression, and attention deficit hyperactivity disorder (ADHD), among other mental illnesses [27]. In addition, it is used for other diseases such as psoriasis, asthma, cystic fibrosis, and diabetes as an adjuvant nutritional supplement in treatment. EPA helps in various kinds of cancer, mostly assisting in body weight maintenance and lessening the negative effects of chemotherapy in cancer patients [28]. When used by women, it lessens the risk of delayed fetal development, high blood pressure during high-risk pregnancy, and menopause symptoms [29]. In fish oil preparations, EPA and DHA are frequently combined for several disorders, such as avoiding heart disease and lowering irregular heartbeats. The idea that encapsulated EPA and DHA supplements improve cardiovascular outcomes in people with known cardiovascular disease or at risk of developing it has been tested over the course of the last 20 years in a number of clinical trials. Nevertheless, using EPA and DHA combination capsules to lower cardiovascular risk is not advised based on the information that is currently available. According to certain research, using EPA on its own may reduce the risk of cardiovascular disease [30].

According to our review study, EPA is used as a supplement after surgery to lessen infections, accelerate wound healing, and speed up recovery [31]. It has been demonstrated that fish oil supplements accelerate the rates at which wounds epithelialize and that omega-3 fatty acids change the inflammatory composition of wounds. It has been demonstrated that EPA and DHA modulate the rate at which wound healing proceeds. The physiology of the skin is changed at the cellular and molecular levels by these distinctive omega-3 fatty acids and their metabolites [30]. On the other hand, there is some controversial data in the literature regarding the oral supplantation of omega-3-fatty acids [30]. It has been observed that oral consumption of omega-3 fatty acids, particularly the DHA and EPA combination, significantly slow down the healing of wounds and disrupt the structure of collagen through several mechanisms [5]. Omega-3 fatty acids work through a variety of mechanisms, such as modifications to the structure of membranes, the synthesis of lipid mediators, and the activation of intracellular receptors that modify the expression of genes. Omega-3 fatty acids stimulate the peroxisome proliferator-activated receptors (PPARs) and therefore could have negative impact on the healing of diabetic ulcers [23,25]. Characteristic genes of mast cells in the skin tissues of patients with type 2 diabetes mellitus (T2DM) were described at a single-cell level. These genes and enriched signaling pathways provide a theoretical basis and data support for further research on dermatopathy in patients with diabetes mellitus [32]. Although the topical application of omega-3 fatty acids could improve wound healing, further research is needed regarding the influence of oral supplementation, particularly on diabetic ulcers.

A number of additional cutting-edge technologies are hitting the market. Among these are products such as Woundchek (DEN180014) (Systagenix) for detecting elevated protease activity as a proxy for impaired wounds; Cellutome (KCI) and Recell (BP170122) (Avita) for epidermal harvest and suspension systems; SofPulse (K070541) (Endonovo) for targeted pulsed electromagnetic therapy; TWO2 (WoundSource) for topical wound oxygen therapy; and UltraMIST (K1407828) (WoundSource) for ultrasound therapy. Numerous growth factors, including platelet-derived growth factor (PDGF), fibroblast growth factor, transforming growth factor-β, and epidermal growth factor, have been identified in the intricate environment of wound healing. Growth factors such as PDGF supplementation [Regranex (BLA103691) (Smith + Nephew)] have been developed as adjuncts treating persistent wounds, such as diabetic neuropathic ulcers, as a result of ongoing technological advancement [2].

However, there are still a lot of obstacles to overcome in the treatment of both acute and chronic wounds, even with the creation of numerous new products. Fundamental knowledge of the pathophysiological mechanisms behind injury and the enhanced identification and comprehension of unmet clinical requirements are necessary. These discoveries will ultimately result in personalized medicine for the treatment of chronic wounds and will enable scientists to determine the optimal application for products with single molecular targets.

Author Contributions: Conceptualization, P.P.; software, M.D. and P.P.; investigation, M.D. and P.P.; data curation, P.P., A.V. and S.P.; writing—original draft preparation, M.D. and P.P.; writing—review and editing, P.P., A.V. and S.P.; supervision, A.V. and P.P. All authors have read and agreed to the published version of the manuscript.

Funding: This research received no external funding.

Conflicts of Interest: The authors declare no conflicts of interest.

References

1. Weller, C.D.; Team, V.; Sussman, G. First-Line Interactive Wound Dressing Update: A Comprehensive Review of the Evidence. *Front. Pharmacol.* **2020**, *11*, 155. [CrossRef] [PubMed] [PubMed Central]
2. Freedman, B.R.; Hwang, C.; Talbot, S.; Hibler, B.; Matoori, S.; Mooney, D.J. Breakthrough treatments for accelerated wound healing. *Sci. Adv.* **2023**, *9*, eade7007. [CrossRef] [PubMed]
3. Yang, P.; Ju, Y.; Liu, X.; Li, Z.; Liu, H.; Yang, M.; Chen, X.; Lei, L.; Fang, B. Natural self-healing injectable hydrogels loaded with exosomes and berberine for infected wound healing. *Mater. Today Bio* **2023**, *23*, 100875. [CrossRef] [PubMed]
4. Huang, T.-H.; Wang, P.-W.; Yang, S.-C.; Chou, W.-L.; Fang, J.-Y. Cosmetic and Therapeutic Applications of Fish Oil's Fatty Acids on the Skin. *Mar. Drugs* **2018**, *16*, 256. [CrossRef] [PubMed]
5. Kiecolt-Glaser, J.K.; Glaser, R.; Christian, L.M. Omega-3 fatty acids and stress-induced immune dysregulation: Implications for wound healing. *Mil. Med.* **2014**, *179* (Suppl. S11), 129–133. [CrossRef] [PubMed]
6. Díaz-García, D.; Filipová, A.; Garza-Veloz, I.; Martinez-Fierro, M.L. A Beginner's Introduction to Skin Stem Cells and Wound Healing. *Int. J. Mol. Sci.* **2021**, *22*, 11030. [CrossRef] [PubMed] [PubMed Central]
7. Kolimi, P.; Narala, S.; Nyavanandi, D.; Youssef, A.A.A.; Dudhipala, N. Innovative Treatment Strategies to Accelerate Wound Healing: Trajectory and Recent Advancements. *Cells* **2022**, *11*, 2439. [CrossRef]
8. Rodrigues, M.; Kosaric, N.; Bonham, C.A.; Gurtner, G.C. Wound Healing: A Cellular Perspective. *Physiol. Rev.* **2019**, *99*, 665–706. [CrossRef] [PubMed] [PubMed Central]
9. Wallace, H.A.; Basehore, B.M.; Zito, P.M. Wound Healing Phases. In *StatPearls*; StatPearls Publishing: Treasure Island, FL, USA, 2024. Available online: https://www.ncbi.nlm.nih.gov/books/NBK470443/ (accessed on 24 June 2024).
10. Sorg, H.; Tilkorn, D.J.; Hager, S.; Hauser, J.; Mirastschijski, U. Skin Wound Healing: An Update on the Current Knowledge and Concepts. *Eur. Surg. Res.* **2017**, *58*, 81–94. [CrossRef] [PubMed]
11. Saini, R.K.; Keum, Y.S. Omega-3 and omega-6 polyunsaturated fatty acids: Dietary sources, metabolism, and significance-A review. *Life Sci.* **2018**, *203*, 255–267. [CrossRef] [PubMed]
12. National Center for Biotechnology Information. PubChem Compound Summary for CID 5282847, Eicosapentaenoic Acid. 2024. Available online: https://pubchem.ncbi.nlm.nih.gov/compound/Eicosapentaenoic-Acid (accessed on 24 June 2024).
13. Rupp, H.; Wagner, D.; Rupp, T.; Schulte, L.M.; Maisch, B. Risk stratification by the "EPA+DHA level" and the "EPA/AA ratio" focus on anti-inflammatory and antiarrhythmogenic effects of long-chain omega-3 fatty acids. *Herz* **2004**, *29*, 673–685. [CrossRef] [PubMed]
14. Sofrona, E.; Tziveleka, L.A.; Harizani, M.; Koroli, P.; Sfiniadakis, I.; Roussis, V.; Rallis, M.; Ioannou, E. In Vivo Evaluation of the Wound Healing Activity of Extracts and Bioactive Constituents of the Marine Isopod *Ceratothoa oestroides*. *Mar. Drugs* **2020**, *18*, 219. [CrossRef] [PubMed] [PubMed Central]
15. Vitsos, A.; Ieronymaki, D.; Kostaki, M.; Almpani, C.; Barda, C.; Kikionis, S.; Sfiniadakis, I.; Dallas, P.; Rallis, M.C. In Vivo Evaluation of Wound Healing Efficacy of Gel-Based Dressings Loaded with Pycnogenol™ and *Ceratothoa oestroides* Extracts. *Gels* **2024**, *10*, 233. [CrossRef]
16. Karoud, W.; Ghlissi, Z.; Krichen, F.; Kallel, R.; Bougatef, H.; Zarai, Z.; Boudawara, T.; Sahnoun, Z.; Sila, A.; Bougatef, A. Oil from hake (*Merluccius merluccius*): Characterization, antioxidant activity, wound healing and anti-inflammatory effects. *J. Tissue Viability* **2020**, *29*, 138–147. [CrossRef] [PubMed]
17. Ontoria-Oviedo, I.; Amaro-Prellezo, E.; Castellano, D.; Venegas-Venegas, E.; González-Santos, F.; Ruiz-Saurí, A.; Pelacho, B.; Prósper, F.; Pérez Del Caz, M.D.; Sepúlveda, P. Topical Administration of a Marine Oil Rich in Pro-Resolving Lipid Mediators Accelerates Wound Healing in Diabetic db/db Mice through Angiogenesis and Macrophage Polarization. *Int. J. Mol. Sci.* **2022**, *23*, 9918. [CrossRef] [PubMed] [PubMed Central]
18. Kotronoulas, A.; de Lomana, A.L.G.; Karvelsson, S.T.; Heijink, M.; Stone Ii, R.; Giera, M.; Rolfsson, O. Lipid mediator profiles of burn wound healing: Acellular cod fish skin grafts promote the formation of EPA and DHA derived lipid mediators following seven days of treatment. *Prostaglandins Leukot. Essent. Fat. Acids* **2021**, *175*, 102358. [CrossRef] [PubMed]
19. McDaniel, J.C.; Szalacha, L.; Sales, M.; Roy, S.; Chafee, S.; Parinandi, N. EPA + DHA supplementation reduces PMN activation in microenvironment of chronic venous leg ulcers: A randomized, double-blind, controlled study. *Wound Repair Regen.* **2017**, *25*, 680–690. [CrossRef] [PubMed] [PubMed Central]
20. McDaniel, J.C.; Rausch, J.; Tan, A. Impact of omega-3 fatty acid oral therapy on healing of chronic venous leg ulcers in older adults: Study protocol for a randomized controlled single-center trial. *Trials* **2020**, *21*, 93. [CrossRef] [PubMed] [PubMed Central]
21. Soleimani, Z.; Hashemdokht, F.; Bahmani, F.; Taghizadeh, M.; Memarzadeh, M.R.; Asemi, Z. Clinical and metabolic response to flaxseed oil omega-3 fatty acids supplementation in patients with diabetic foot ulcer: A randomized, double-blind, placebo-controlled trial. *J. Diabetes Complicat.* **2017**, *31*, 1394–1400. [CrossRef] [PubMed]
22. Babaei, S.; Ansarihadipour, H.; Nakhaei, M.; Darabi, M.; Bayat, P.; Sakhaei, M.; Baazm, M.; Mohammadhoseiny, A. Effect of Omegaven on mast cell concentration in diabetic wound healing. *J. Tissue Viability* **2017**, *26*, 125–130. [CrossRef] [PubMed]
23. Burger, B.; Kühl, C.M.C.; Candreva, T.; Cardoso, R.D.S.; Silva, J.R.; Castelucci, B.G.; Consonni, S.R.; Fisk, H.L.; Calder, P.C.; Vinolo, M.A.R.; et al. Oral administration of EPA-rich oil impairs collagen reorganization due to elevated production of IL-10 during skin wound healing in mice. *Sci. Rep.* **2019**, *9*, 9119. [CrossRef] [PubMed] [PubMed Central]

24. Candreva, T.; Kühl, C.M.C.; Burger, B.; Dos Anjos, M.B.P.; Torsoni, M.A.; Consonni, S.R.; Crisma, A.R.; Fisk, H.L.; Calder, P.C.; de Mato, F.C.P.; et al. Docosahexaenoic acid slows inflammation resolution and impairs the quality of healed skin tissue. *Clin. Sci.* **2019**, *133*, 2345–2360. [CrossRef] [PubMed]
25. Burger, B.; Sagiorato, R.N.; Silva, J.R.; Candreva, T.; Pacheco, M.R.; White, D.; Castelucci, B.G.; Pral, L.P.; Fisk, H.L.; Rabelo, I.L.A.; et al. Eicosapentaenoic acid-rich oil supplementation activates PPAR-γ and delays skin wound healing in type 1 diabetic mice. *Front. Immunol.* **2023**, *14*, 1141731. [CrossRef] [PubMed] [PubMed Central]
26. Bhatt, D.L.; Steg, P.G.; Miller, M.; Brinton, E.A.; Jacobson, T.A.; Ketchum, S.B.; Doyle, R.T., Jr.; Juliano, R.A.; Jiao, L.; Granowitz, C.; et al. Cardiovascular Risk Reduction with Icosapent Ethyl for Hypertriglyceridemia. *N. Engl. J. Med.* **2019**, *380*, 11–22. [CrossRef] [PubMed]
27. Chang, J.P.; Su, K.P.; Mondelli, V.; Satyanarayanan, S.K.; Yang, H.T.; Chiang, Y.J.; Chen, H.T.; Pariante, C.M. High-dose eicosapentaenoic acid (EPA) improves attention and vigilance in children and adolescents with attention deficit hyperactivity disorder (ADHD) and low endogenous EPA levels. *Transl. Psychiatry* **2019**, *9*, 303. [CrossRef] [PubMed] [PubMed Central]
28. Paixão, E.M.D.S.; Oliveira, A.C.M.; Pizato, N.; Muniz-Junqueira, M.I.; Magalhães, K.G.; Nakano, E.Y.; Ito, M.K. The effects of EPA and DHA enriched fish oil on nutritional and immunological markers of treatment naïve breast cancer patients: A randomized double-blind controlled trial. *Nutr. J.* **2017**, *16*, 71. [CrossRef] [PubMed] [PubMed Central]
29. Natural Medicines Comprehensive Database (NMCD), Therapeutic Research Faculty. Eicosapentaenoic Acid (EPA): Overview, Uses, Side Effects, Precautions, Interactions, Dosing and Reviews. 2018. Available online: https://www.webmd.com/vitamins/ai/ingredientmono-994/eicosapentaenoic-acid-epa (accessed on 24 June 2024).
30. Khoukaz, H.B.; Fay, W.P. Fish Oil Supplements for Prevention of Cardiovascular Disease: The Jury Is Still Out: CON: Fish Oil Is Useful to Prevent or Treat Cardiovascular Disease. *Mo. Med.* **2021**, *118*, 219–225. [PubMed]
31. Seth, N.; Chopra, D.; Lev-Tov, H. Fish Skin Grafts with Omega-3 for Treatment of Chronic Wounds: Exploring the Role of Omega-3 Fatty Acids in Wound Healing and A Review of Clinical Healing Outcomes. *Surg. Technol. Int.* **2022**, *40*, 38–46. [PubMed]
32. Liao, B.; Ouyang, Q.; Song, H.; Wang, Z.; Ou, J.; Huang, J.; Liu, L. The transcriptional characteristics of mast cells derived from skin tissue in type 2 diabetes patients at the single-cell level. *Acta Histochem.* **2021**, *123*, 151789. [CrossRef]

www.ingramcontent.com/pod-product-compliance
Lightning Source LLC
Chambersburg PA
CBHW062045220926
51850CB00010B/2668
* 9 7 8 3 7 2 5 8 5 9 2 9 0 *

MDPI AG
Grosspeteranlage 5
4052 Basel
Switzerland
Tel.: +41 61 683 77 34

Applied Sciences Editorial Office
E-mail: applsci@mdpi.com
www.mdpi.com/journal/applsci